Meteorites

Their Impact on Science and History

Edited by
Brigitte Zanda and Monica Rotaru
Translated by Roger Hewins

CAMBRIDGE
UNIVERSITY PRESS

PUBLISHED BY THE PRESS SYNDICATE OF THE UNIVERSITY OF CAMBRIDGE
The Pitt Building, Trumpington Street, Cambridge, United Kingdom

CAMBRIDGE UNIVERSITY PRESS
The Edinburgh Building, Cambridge CB2 2RU, UK
40 West 20th Street, New York, NY 10011-4211, USA
10 Stamford Road, Oakleigh, VIC 3166, Australia
Ruiz de Alarcón 13, 28014 Madrid, Spain
Dock House, The Waterfront, Cape Town 8001, South Africa

http://www.cambridge.org

Originally published in French as *Les Météorites* © Bordas, Paris 1996
English edition © Cambridge University Press 2001

First published 2001

Printed in Italy

Typefaces Frutiger and Bodoni   *System* QuarkXPress©

A catalogue record for this book is available from the British Library

Library of Congress Cataloguing in Publication data available

ISBN 0 521 79940 6 paperback

The following contributed to the preparation of this book:

under the direction of Brigitte Zanda and Monica Rotaru,
assisted by Philippe de la Cotardière,

Daniel Benest, Jean-Pierre Bibring, Michèle Bourot Denise, Michel Festou, Roger Hewins,
Catherine Caillet Komorowski, Hugues Leroux, Mireille Christophe Michel Lévy, George McGhee,
Gérard Manhès, Ursula Marvin, Kenneth Miller, Paul Pellas, Claude Perron, François Raulin,
François Robert, Eric Robin, Robert Rocchia, Pierre Thomas, Ernst Zinner.

For the exposition "Météorites!", the Muséum National d'Histoire Naturelle wanted to present to the public a large part of the national collection of meteorites for which it has responsibility. This collection was inaugurated in 1861 at the instigation of A. Daubrée, Professor of Geology, by combining the fourteen specimens curated in the Mineralogy Laboratory with the sixty four which existed in the Geology collection. It grew rapidly, to 268 specimens in 1878, and soon became one of the leaders among the great meteorite collections of the world. A. Lacroix, Professor of Mineralogy, arranged in 1926 that the collection of meteorites, which then contained 570 specimens, be transferred from the Geology Laboratory to the Mineralogy Laboratory.

While the Museum's collection remained world class until the end of the Second World War, with 670 specimens in 1950, it has not progressed since like the majority of great collections, because of the insufficiency of funding for acquisitions, and because of the absence of general rules for recuperation of météorites on the national territory. As early as 1909, S. Meunier, Professor of Geology, denounced these difficulties: "It is essential to note that our national collection, however beautiful and interesting it may be, is not as extensive as it could be. Specimens are offered to the Museum that it is impossible to acquire because of the very considerable price that they have commanded in recent years. In spite of everything, the collection has grown in a very honourable manner, reaching 974 specimens in 1996 and 1145 in 2000.

In parallel with our exposition, the Museum wanted to publish a work bringing up to date the knowledge acquired by the study of this extraterrestrial material. This book presents the principal results of international research, as demonstrated by the participation of several foreign colleagues alongside French workers, all specialists in this domain. In the name of the Muséum National d'Histoire Naturelle, I very warmly thank all the authors for their kind collaboration.

Jacques Fabriès
Former Professor at the Muséum National D'Histoire Naturelle,
Former Director of the Mineralogy Laboratory

CONTENTS

hey fall

For a few seconds, for the spectators who are packed in the stands of football stadiums in the north-east of the United States, on the evening of October 9th 1992, the show is no longer on the field, it is in the sky: a luminous object, brighter than the full moon, has just appeared above their heads.

It moves at a great speed towards the north-east on a nearly horizontal trajectory. Suddenly the fireball splits into several tens of fragments, creating a long trail which finishes by disappearing after having flown 700 km, from Kentucky to New Jersey. A few instants later, Michelle Knapp rushes to the window of the house she lives in, in the little town of Peekskill in New York state, after hearing a violent noise. The trunk of her car, parked under her windows, is deformed and pierced by a hole. She calls the police. After a brief inspection, they discover driven into the asphalt under the car, a 12 kg black stone, as big as an American football. It's a neighbour who lets out the word – meteorite!

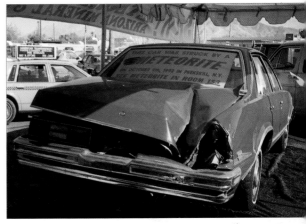

Michelle Knapp's car still bears the mark of the arrival of the Peekskill meteorite. It has become a museum piece.

METEORS AND METEORITES

The unusual spectacle that thousands of Americans have just witnessed is indeed a meteorite fall. The luminous phenomenon, the meteor, was due to the intense heat loss which accompanied the slowing of the

The Peekskill meteor. More than 70 fragments can be distinguished, of which only one was found on the ground.

Left page: shooting star. This familiar brief light signals the arrival of cosmic dust grain at the top of the atmosphere.

object in the air. In space, the meteorite orbited the Sun at a speed, a little before the encounter with the Earth, of 40 km/s. Our planet orbiting itself in the same direction at 30 km/s, the speed relative to the Earth was only 10 km/s, which the acceleration due to

Passage through the atmosphere

The spectacular character of meteorite falls is linked to the presence of the terrestrial atmosphere. At the very great speeds (15 to 20 km/s) with which these extraterrestrial bodies arrive, friction with the air can bring their surface to several thousand degrees: this is the origin of the luminous phenomenon (meteors). The smaller the mass of the meteorite, the greater the braking action. Dust is stopped in the highest layers of the atmosphere and the corresponding meteor lasts only a fraction of a second: this is what we call a shooting star.

Greater masses penetrate deeper into the atmosphere and cause phenomena observable for longer times. The meteor lights up at an altitude of a hundred kilometres. The surface material of the projectile is melted, vaporised and ejected, leaving in its wake a trail of ionised matter, vapour and dust, which can last several minutes after the fall. The mass of the meteorite thus decreases continuously: this is the phenomenon of ablation. The terrible tensions to which it is liable very often provoke its fragmentation into two or several pieces, even thousands. Near an altitude of 20 km, the latter, if they have survived, no longer have a speed sufficient to maintain incandescence. The meteor goes out. The trajectory of the remaining fragments bends little by little towards the ground and they finish their flight in free fall, distributing themselves over the ground in a zone called the strewn field, which can extend for tens of kilometres. Stony meteorites about 10 to 100 m in diameter explode in the atmosphere, the bigger ones lower down. Larger stones and irons reach right to the ground. They are very little slowed down and hit the surface with a speed close to their cosmic velocity. The shock is then very violent and an impact crater is formed. Sound phenomena frequently accompany the fall of a meteorite. The shock wave which grows in front of the fireball is felt like the "bang" of a plane crossing the sound barrier, and the air turbulence which is produced behind it can cause noise like the rumble of thunder, which reaches the ground after the fall of the meteorite because it is propagated at a much lower speed .

the gravitational attraction of the Earth brought to 15 km/s, or 54,000 km/h, at the top of the atmosphere. This is a typical speed for a meteorite. We understand easily that the effects of passing through the atmosphere at such a speed are terrible. If the stone found under Michele Knapp is so black, if all its angles are rounded, giving it the characteristic aspect of meteorites, it is because of this violent passage. Once the stone is broken, we see that only a thin external skin is black. This is what we call the fusion crust. It is exceptionally thick in the case of Peekskill (1 mm), but usually does not exceed a few tenths of a millimetre. This skin has been melted, it is glassy, but the interior of the meteorite is intact for two reasons: on the one hand the external part of

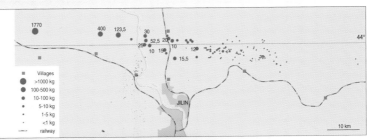

Strewn field of the Jilin meteorite which fell in China on March 8th 1976. The mass of the fragments is indicated by the size of the points.

Villages
>1000 kg
100-500 kg
10-100 kg
5-10 kg
1-5 kg
<1 kg
railway

10 km

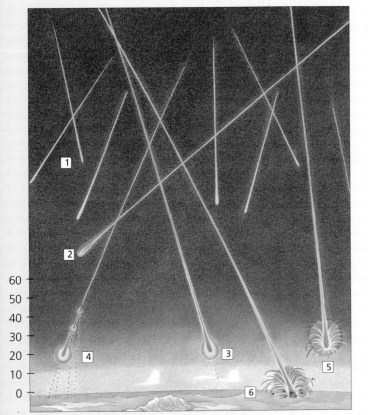

Schematic representation of the atmospheric passage of different meteorites.
In order of increasing masses:
1 shooting stars.
2 fireball; the object is entirely destroyed by ablation in the atmosphere.
3 the object is mostly destroyed in the atmosphere and a meteorite reaches the ground.
4 fragmentation of a meteorite whose pieces spread over a fall ellipse.
5 stony meteorite of about 10 metres in diameter exploding in the atmosphere.
6 very big meteorite forming an impact crater on reaching the ground.

the meteorite, which has been melted or volatilised and lost in the atmosphere during the fall, has carried away a great part of the heat produced; on the other hand, the incandescence only lasts a few seconds, which does not leave time for the heat to diffuse deeply (there again, Peekskill is unique with a meteor lasting 40 s, because of its almost horizontal trajectory). If we exclude the first few millimetres near the fusion crust, which can be slightly affected, the interiors of meteorites are thus not affected by their fall, which offers us a chance to decode the precious information they contain.

CAN THE SKY FALL ON OUR HEADS?

If hundreds of millions of cosmic dust particles, responsible for shooting stars, arrive every second at the top of the terrestrial atmosphere, we estimate that for meteorites of more than 1 kg the corresponding flux is only about 100,000 per year for the entire Earth. How many of these reach the surface of our planet? It is difficult to determine, for many are entirely destroyed during their passage through the atmosphere. It seems that several thousands of meteorites greater than 1 kg reach the surface of the Earth each year, of which about a hundred are greater than 100 kg. Only a small fraction of the latter are recuperated. It has very rarely happened – at least during historical times – that a meteorite fall caused serious damage. The Peekskill car is only the second to have been hit this way. It has happened several times the roof of a house was pierced by a meteorite, a boat was hit in Japan in March 1991, and it is said that a dog was killed by the fall of the Nakhla meteorite in Egypt on June 28th 1911. On August 14th 1992, a rain of meteorites crashed down on Mbale, in Uganda, causing minor damage to a few houses. More than 800 fragments were found. One of them fell on the head of a young boy, doing

Slice of the Peekskill meteorite. The violence of the atmospheric passage has not affected the interior of the meteorite which remains perfectly intact.

him no harm; it only weighed 3.6 g and its fall was broken by the leaves of a banana tree. Very happily for the inhabitants of the Earth, the biggest meteorites are even more rare. It is estimated that a meteoritic projectile 10 km in diameter (about a thousand billion tons) reaches the Earth every 100 million years. The consequences of such a fall would, obviously, be dramatic.

A CATASTROPHE AVOIDED

A rather special "fall" deserves a note. On August 10th 1972, at the beginning of the afternoon, thousands of inhabitants of western United States and Canada saw above their heads the passage of a superb fireball, running from

Tamentit iron meteorite. Its surface has been sculpted by the atmospheric passage, causing the typical regmaglypts.

Ouallen meteorite, found by Théodore Monod in the Sahara in 1936. The fine black skin which covers it is the fusion crust, formed during passage through the atmosphere.

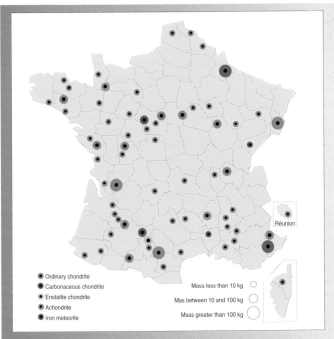

Sites of 68 of the 71 meteorites located on French territory in the last 500 years.

Réunion

- Ordinary chondrite
- Carbonaceous chondrite
- Enstatite chondrite
- Achondrite
- Iron meteorite

Mass less than 10 kg ○
Mas between 10 and 100 kg ◯
Mass greater than 100 kg ◯

The meteorites we collect

Two thirds of the surface of the globe being covered with oceans, two thirds of the meteorites are lost. Only a small fraction of the others can be collected immediately after their fall – 6 meteorites per year, on average, during the twentieth century. Meteorites which no one sees fall are slowly altered. They are however better preserved in the dry climates of hot or cold deserts than in humid or temperate regions. Some can thus be spotted long after their fall: these are what we call finds (as opposed to falls). In total, in 2000, we know 4745 non-Antarctic meteorites (1147 observed falls) and over 18,000 fragments of Antarctic meteorites, coming from about 5,000 different meteorites, but about 600 hot desert and 2,000 Antarctic finds are waiting to be classified and published. In France, since 1492 (fall of the Ensisheim meteorite), they have recorded 71 meteorites of which 61 are falls and 10 finds (including one from the Island of Réunion). The meteorites are named after the place closest to the spot where they were collected. For Antarctic meteorites, an acronym which designates the collection zone is followed by numbers indicating the year and sample number.

the south to the north. Some heard a violent explosion, but nobody witnessed the fall of the meteorite. After grazing the Earth, it did indeed set off again into space, where it continued its course. The best data on the phenomenon come from an infrared detector, launched on an American military satellite. This detector spotted the meteorite at an altitude of 76 km, and lost it 1 min 40 s later, 1,500 km further north, at an

The fireball of August 10th 1972, photographed above Jackson Hole, Wyoming, when it was at an altitude of about 60 km and moving at 15 km/s. It returned to space and has not been seen since.

altitude of 102 km. In between, the fireball had approached the ground by as little as 58 km. The size and mass of this object are plausibly of the order of 5 m and 100 tons. It should have come back to the vicinity of the Earth in August 1997, but was not seen.

THE EARTH IS GROWING

What is the mass deposited on the Earth in 4 billion years? We estimate it today as 100,000 or 200,000 tons per year on average. Essentially 3 types of object contribute to this influx, in a roughly equivalent manner. The first contribution comes from very big meteorites, 1 to 10 km in diameter. However, because of their rarity, their yield is significant only when considered over very long times (hundreds

of millions of years). In the course of the last few thousand years, their contribution has been nil. The second contribution is that of objects 10 to 50 m in diameter, which we have perceived recently, by telescope observations, to be much more frequent than we supposed. Several of them probably arrive on Earth per year. The third contribution is that of dust particles with a size between 0.05 and 0.5 mm. It is about 40,000 tons per year (100 tons per day). Meteorites of normal size contribute relatively little to the capture of cosmic material by the Earth.

AN INFINITESIMAL INFLUX

What does this cosmic influx represent in comparison with the total mass of the Earth? It is easy to calculate that, in the

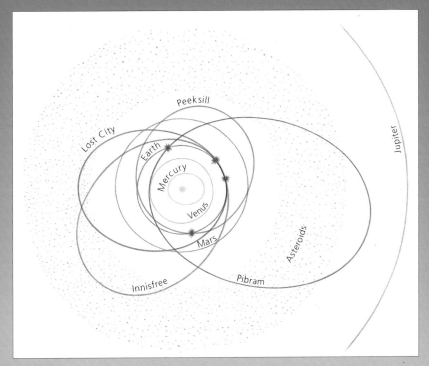

Right: orbits of four meteorites whose fall was photographed or filmed, as well as the inner planets. The location of the asteroid belt is indicated.

Below: fall of the Lost City meteorite, photographed above Oklahoma on January 3rd 1970 by the U.S. photographic network. The meteor remained visible for 9 seconds. The parallel traces are due to the movements of the stars during the four and a half hour exposure.

Photographic networks

From 1964 on, vast networks of cameras pointed every night towards the sky, were installed in Czechoslovakia, the United States, and then Canada, to record meteorite trajectories and help locate possible meteorites. A shutter closes the lens at regular intervals for very short times, which allows the determination of the speed of the fireball from the corresponding marks left on the photos. If a meteor is photographed by several cameras installed in different places, we can then, by triangulation, reconstitute the trajectory in the atmosphere and in space. This also permits, if it seems that a meteorite has reached the ground, a quite precise determination of its fall zone. During the twenty years that they functioned, the three networks made possible the recovery of one meteorite each. If we add to that the Peekskill meteorite, filmed by at least fourteen video cameras, four falls have been recorded, which has allowed us to determine the corresponding orbits with precision.

Exposure ages

Our galaxy is threaded with cosmic rays, formed by atomic nuclei excited to very high energy, like the particles which nuclear physicists propel in their accelerators. We are protected from this radiation by the terrestrial atmosphere, but this is not the case for asteroid surfaces, stripped of atmosphere. However, the most energetic particles only penetrate a few metres into the interior of a solid. After a collision between asteroids, the rocky blocks which are ejected into space, and which become meteorites, are in general exposed for the first time to cosmic rays. They were protected up to then by the rock which covered them (unless they were located at the surface of the asteroid). During all their long interplanetary tour up to the Earth, they will suffer bombardment by cosmic particles. These will induce nuclear reactions in them, that is to say the splitting of atomic nuclei which they will transform into other, lighter atomic nuclei. The abundance of these "cosmogenic" nuclides increases with time if they are stable. Certain cosmogenic nuclei are radioactive: they can be analysed in installations designed to measure weak radioactivity. Among the stable cosmogenic nuclei, those of the rare gasses are the most easily measured. These measurements allow us to determine, for each meteorite, its exposure age, which is to say the time elapsed between its ejection from the parent asteroid and its arrival on Earth.

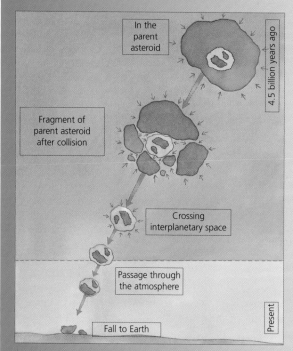

In the parent asteroid

Fragment of parent asteroid after collision

Crossing interplanetary space

Passage through the atmosphere

Fall to Earth

4.5 billion years ago

Present

course of the last 4 billion years, our planet has accreted on the order of 500,000 billion tons of extraterrestrial matter, or a ten millionth of its mass. If this material had stayed uniformly spread over the entire surface of the globe, it would form a layer about 40 cm thick (recall that the radius of the Earth, for comparison, is about 6,400 km). As a result of the movements which affect the various plates making up the upper part of the Earth – plate tectonics – most of this material has been dragged into the depths of the terrestrial globe. These numbers are approximate, but it would be even more chancy to extend the calculation back to the birth of the Earth (about 4.5

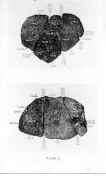

Left, reassembly of the pieces of the St. Severin meteorite and, below, the pre-atmospheric form of the same meteorite.

The meteorite which pierced the roof of a house in Mihonoseki, in Japan, on December 10th 1992, did not stop until two floors lower.

billion years ago), the meteoritic bombardment having been much more intense in the course of the first 500 million years of the planet.

Let's come back for a minute to the meteorites of 10 to 50 m mentioned above. That several per year arrive on Earth seems astonishing. We could indeed think that the fall of objects of this size could not pass unperceived for long. The resolution of this paradox comes from two directions. On the one hand, calculations treating their passage through the atmosphere have shown that stony meteorites (but not the equivalent iron ones) of 10 to 100 m diameter explode in the atmosphere and only reach the ground in the state of dust. Only the biggest ones are destroyed low enough that the effects of the explosion are felt on the ground. This is, perhaps, what happened on June 30th in Siberia, in the region of Tunguska. On the other hand, observations

classed as defence secrets until the end of 1993 show that more than 200 of these explosions have indeed been seen by American spy satellites since 1972.

WHERE DO METEORITES COME FROM?

It is well known that the frequency of shooting stars increases at certain periods of the year, when the Earth, in its movement around the Sun, crosses the trajectory of a comet. Dust is ejected by the comet, which continues to travel on its orbit. A fraction at least of shooting stars thus come from comets. This is not the case with meteorites, which are in the great majority fragments of asteroids. The measurement of their cosmic ray exposure ages shows that the duration of their trip between their parent asteroid and the Earth varies from 1 to 100 million years for stones. It can be up to 2 billion years for iron meteorites, probably reflecting their greater resistance to erosion in space.

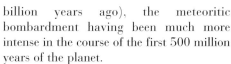

Key words: **ablation • accretion • asteroid • fireball • comet • cosmic** material • **cosmic** dust • **cosmic rays • crater • exposure age • fall • fall ellipse • find • fusion crust • ionised** matter • **meteor • meteorite • nuclear reaction • nucleus • orbit • plate tectonics • regmaglypt • shock wave • shooting star**

ell from the sky

Since time immemorial, a celestial origin has been attributed to exotic masses of iron found on the ground, and they often have been venerated. In 1576, natives of the wild reaches of northern Argentina led Spanish soldiers to a large half-buried iron, and in 1772 inhabitants of Siberia showed a metallic mass to the explorer Pallas: in both cases the local peoples said that the masses had fallen from the sky. Although they could not convince their visitors, these unlettered men had a better intuition than the explorers did about the origin of these irons.

Above: representation of a fragment of the metal-rich mass shown to Peter Simon Pallas in 1772, and later identified as a meteorite. We call this very rare category of meteorite, consisting of metal studded with large olivine crystals, "pallasites.".

THE ENSISHEIM METEORITE, 1492

Recently, a meteorite that fell into a monastery in Japan in the year 861 was found preserved there. The second observed fall of which we still have material occurred at Ensisheim in Alsace on November 7th, 1492, at 11:30 in the morning. After a gigantic explosion that resounded over the upper Rhineland, a great black stone plunged from the sky into a wheat field just outside the city walls. The only witness to the fall was a young boy who led an excited crowd to a hole 2 m deep with the stone lying in the bottom. They pulled out the stone and estimated its weight at 150 kg. Immediately, people began hacking off

Left page: this painting by Albrecht Dürer is on the back of a small wood panel representing the penitent St. Jerome. Tentatively dated to 1494, it provides convincing evidence that the artist witnessed the final explosion of the Ensisheim fireball while he was living in nearby Basel in November, 1492.

Above: Melancolia I, engraved in 1514 by Albrecht Dürer, shows a fireball explosion, most likely the Ensisheim meteorite.

Jon dem donnerstein gefallē jm rcij.iar:vor Ensißhein

Battenhem

Ensßheim

Defulgetra anni rcij.

O ych wundert mancher frember gſc
Der merck vnd leß ouch diß bert:

*Above: the news of the Ensisheim fall soon
spread and acquired a political significance. In
Basel, the poet Sebastian Brant wrote Latin and
German verses and had them printed on
broadsheets to celebrate the event and to
encourage King Maximilian to attack his French
enemy, Charles VIII. To give the idea of
movement, the stone is shown in the air and on
the ground in this woodcut accompanying the
verses.*

*Right: the fall of the stone represented in 1513 in
the chronicle of Lucerne by Diebold Schilling*

pieces to carry as good luck charms, until
the bailiff appeared and stopped the
destruction. He then ordered the stone
carried into the city and placed in front of
the church door. Nineteen days later,
Maximilian the young king and future
emperor of Austria, stopped at Ensisheim as
he led his army to battle the French. His
advisors declared the stone to be a sign of
divine grace toward the King and a presage
of his victory. Maximilian ordered it to be
preserved in the church as evidence of a
miracle. It remained there for three hundred
years and then was moved to Colmar where
numerous fragments were chipped off for

*View of the Ensisheim stone as it appears today,
rounded by the taking of numerous samples.
Its mass is close to 56 kg and it still has
remnants of fusion crust.*

The Ensisheim fall, good or bad omen?

After the poems of Brant urging King Maximilian to attack his enemies without delay, the King won a battle, thus causing the stone to be confirmed as a blessing from God, at least by the German chroniclers. However, a more pessimistic view appeared in an oil painting inspired by Brant's broadsheets and incorporated by Sigismondo Tizio (1458-1528) in his history of Siena in ten volumes. The stone emerges from a dark cloud revealing a face topped with horns (the wind or the devil?), dead birds fall from the sky, an unidentified animal hides in its burrow, and a salamander (symbol of fire) creeps away from the stone. Finally, an owl, a bird of bad omen, watches over the event. For the Italian chronicler, the stone presaged the election of the Spaniard, Rodrigo Borgia, as Pope Alexander VI, the invasion of Italy by Charles VIII, and the ravages of syphilis in Europe at the beginning of the sixteenth century, which was thought to have been introduced upon the return of Christopher Columbus from the New World.

analysis. In 1803, it was returned to Ensisheim, where today a rounded 56 kg fragment, still showing patches of fusion crust, is exhibited in the museum in the old city hall. .

THE REALITY OF FALLS RECONSIDERED

Until the Renaissance, no one doubted that stones fall from the sky. In a publication

Stones falling from the sky... On this print of a work dating from 1517, a fossil splits a tree in two and a shark's tooth threatens to kill a man...

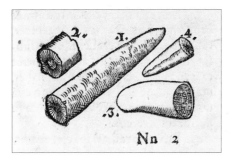

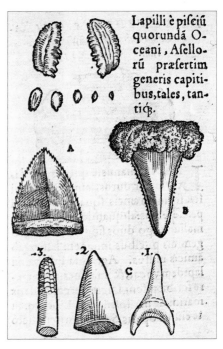

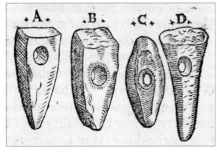

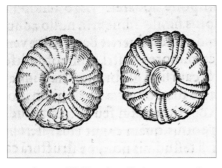

Stones fallen from the sky described by Conrad Gesner (1516–1563). Since ancient times, it was believed that stones from the sky had special forms. In these earliest illustrations, printed in 1565, we recognise fossil urchins, belemnites and prehistoric tools. Triangular objects, which were believed to fall from the sky during lunar eclipses, were fossil shark's teeth.

dating from 1565, the Swiss naturalist, Conrad Gesner, listed as fallen stones some things we recognise today as fossils, mineral concretions, and prehistoric axe heads. However, at the beginning of the eighteenth century, when the real origin of these types of objects had been established, scholars dismissed as vulgar superstition all stories of falling stones and fragments of iron.

THE ANALYSIS OF THE LUCÉ STONE, 1777

The first stone to be analysed by modern chemical methods fell near a group of harvesters at Lucé, on September 13th, 1768. Sent to the Institut de France by the abbot Bachelay, the stone was analysed by three chemists: Fourgeroux de Bonderoy, Cadet de Gassicourt and Antoine-Laurent de Lavoisier, who found it to consist of 55% vitrifiable earth, 6% iron and 8.5% sulphur. They did not consider it to be at all extraordinary. "We believe we can therefore conclude (...) that the stone presented by M. Bachelay did not originate in thunder; that it did not fall from the sky (...), that this stone is nothing other than a pyrite-bearing sandstone (...) which was perhaps covered by a thin layer of soil or grass that was

Top left: iron oxide tumbled in a river; top right: a fragment of an artificial satellite; lower left: a concretion of marcasite; bottom right: furnace slag.

Objects commonly mistaken for meteorites

The chances of finding a true meteorite are extremely rare, unless it falls in daylight close to an observer. Nevertheless, dozens of "meteowrongs" are brought every year to museum curators, who are sometimes told that they were hot when collected or that they singed dry leaves. The curators are very sceptical, because they know that rocks are poor conductors of heat and that fireballs surrounding the falling bodies burn out after only a few seconds. However, they willingly examine these objects in the hope of finding a new meteorite. Unfortunately, most of the specimens are concretions of marcasite or iron oxide, irregular masses of porous limestone, heavy masses of dark coloured magnetite, foundry slags, fine-grained basalt pebbles rolled by the waves or glaciers, or rare fragments of artificial earth-orbiting satellites. Sometimes, even today, they are fossils like those described by Gesner.

struck by lightning, and the heat was great enough to melt the surface of the part struck; but it did not continue long enough to penetrate the interior." (Report published in 1777.) Their examination of a second stone of almost identical composition, which fell near Coutances in Normandy, did not change their minds. "We do not believe that one can conclude anything else from this resemblance, unless it is that lightning strikes pyrite-bearing material preferentially."

THE AGRAM IRON AND THE EICHSTÄDT STONE

In 1790, the abbot Andreas Stütz, of the natural history museum in Vienna (Austria), studied two of the stones in the collection. One of them (an iron mass of about thirty kg) had reportedly been seen to fall at Agram, in Croatia, one evening in 1751. Seven witnesses described a great ball of fire in the sky splitting into two during an enormous explosion. The second presumed fall was that of a little black stone recovered after a loud explosion in February, 1785, near Eichstädt in Bavaria.

The sincerity with which these events had been described and the similarities of the accounts convinced Stütz that something unknown but very real lay behind these two events. He was not referring to possible falls from the sky, however. Recent writings about electricity and thunder convinced Stütz that lightning striking ordinary rocks could reduce iron oxide to metal, and simultaneously would confuse witnesses into believing they had seen the rocks fall from the sky. The radically different interpretation of the same facts by the German physicist Chladni, however, would soon open a new era.

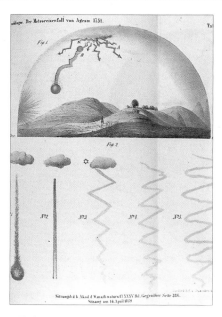

The first modern print of a meteoritic fireball: Agram, 26 May 1751. Top, two fireballs and lightning flashes emerge from a black cloud. Bottom, a fireball falls to the Earth followed by a smoky trail that fades when the first evening star appears. The cloud disappears later and the long trail, windblown into a zigzag shape, remains visible until nightfall.

THE SHOWER OF STONES AT SIENA, 1794

On June 16th 1794, about 7 p.m., an exceptionally high, dark cloud approached Siena from the north, emitting smoke, sparks, and flashes of red lightning. After an ear-splitting explosion, stones fell at the feet of such a multitude of witnesses that it became impossible to deny the reality of the fall or to pretend that it had been observed only by ignorant people. Before the end of the

One of the stones from the fall of Siena, in Italy, June 16th 1794 at 7 p.m.

Chladni's hypotheses, 1794

The German physicist Ernst Florens Friedrich Chladni published a little book in 1794, proposing that stones and masses of iron actually do fall from the sky. He argued that the similarities in the testimony of eyewitnesses from place to place and from century to century supported the explosion of fireballs and the fall of fragments. He added that masses of stone and iron originate in space and give rise to the observed fireballs due to the heat caused by friction with the terrestrial atmosphere. Chladni's hypotheses violated all the best established scientific dogmas of the period. Had not Isaac Newton himself declared that small bodies could not exist in space beyond the Moon? Therefore, there existed no sources from which the masses could come, unless they formed high in the atmosphere by the coagulation of dust under the influence of lightning or of jets of burning gas arising from the Earth. Chladni's German readers rejected his ideas, and almost two years passed before his book was distributed elsewhere in Europe. It took nearly a decade for the idea of falls to be accepted, and much longer for his hypotheses of origin in space and a link with fireballs to take hold.

year, two eminent Italian scholars published well-documented accounts of the event, thus raising the subject of fallen stones to the level of scientific discourse. The Siena fall was mentioned in a report by Sir William Hamilton on the simultaneous eruption of Mt. Vesuvius that was published in England, and helped to prepare the way, a year in advance, for spreading the theories of Chladni in this country.

THE ANALYSES OF HOWARD, DE BOURNON AND VAUQUELIN

The next witnessed fall of a stone occurred on December 13th 1795, at Wold Cottage in England. By a happy chance, a piece of the stone came into the hands of Sir Joseph Banks, president of the Royal Society, who recognised the resemblance between this stone, partially covered with a black crust, and one he possessed from the fall at Siena. Subsequent reports of new falls in Portugal in 1796, and in India in 1798 persuaded Sir Joseph that it was time for a serious study of this phenomenon. He gave his samples of the Siena and Wold Cottage stones to the young chemist, Edward C. Howard, who acquired two more fallen stones and fragments of four so-called "native irons". These included the iron the Spanish had sampled at Campo del Cielo, Argentina, and the one from Siberia, which Pallas had shipped to Germany. Howard obtained the assistance of the French émigré mineralogist, Jacques-Louis de Bournon, who separated each stone into its four main

The Wold Cottage fall provided proofs supporting the theories of Chladni. In 1799, the owner of the property had a commemorative monument erected at the place of fall on which one can read: " On this spot, December 13th 1795, fell from the Atmosphere AN EXTRAORDINARY STONE (...) THIS COLUMN In Memory of it was erected by EDWARD TOPHAM 1799."

components: magnetic grains of metal, reddish iron sulphides, "curious globules," and a fine-grained earthy matrix. Howard analysed each of these constituents separately and found striking similarities in the mineralogy, texture, and chemical composition of the four stones. His most important discovery was of a significant quantity of nickel in all the irons and the metal grains of the stones. This linked the stones with irons and decisively set "fallen bodies" apart from rocks of the Earth's crust.

In 1802 and 1803, the results of Howard and Bournon were presented to the Royal Society and then to the Institut de France, which also heard those of the chemist Nicolas Louis Vauquelin whose conclusions were similar. Several of the leading scientists of the time, including Laplace, Poisson, and Biot, were thus convinced not only that stones could fall from the sky, but also that their origin had to be sought outside the Earth – probably in the volcanoes of the Moon.

THE FALL AT L'AIGLE, 1803

The last doubts on this subject were swept away at 1:00 p.m. on April 26th 1803, when a shower of nearly three thousand stones fell near the community of L'Aigle in Normandy. In the following days, while specimens were hawked in the streets of Paris, Vauquelin and Fourcroy analysed samples and found them to contain nickel and to match earlier fallen stones in other respects. Biot's report, published by the Institut de France in 1803, put an end to the controversy and brought honour to Chladni for his farsightedness. But decades would pass before the link between falls and fireballs would be established, and it would take more than a century and a half for meteorites to be recognised as impact debris from collisions of asteroids with one another and with the Moon and Mars.

A few specimens of the fall of L'Aigle.

Jean-Baptiste Biot's report on the fall of L'Aigle

Jean-Baptiste Biot was sent by Chaptal, the minister of the interior, to study in the field the nature and extent of this remarkable fall. He approached L'Aigle by a roundabout route, questioning witnesses along the way. The first map ever published of a meteorite strewn field appeared in his report of 1803 to the Institut de France. "The meteor did not burst exactly at L'Aigle, but half a league from there (...) before the explosion of the 6th of Floréal nobody ever saw meteoritic stones in the hands of the inhabitants of the region (...) the foundries, the factories and the mines of the region, which I visited, exhibited nothing among their products or their slags which has the slightest relationship with these substances. We do not see in the region any trace of a volcano... I leave to the cleverness of physicists the numerous consequences that one can deduce from this, and I shall count myself happy if I have succeeded in putting one of the most astonishing phenomena that man has ever observed beyond doubt." (Text extracted from the report of J.-B. Biot to Chaptal.)

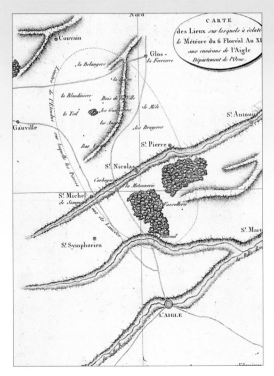

The first map ever published of a meteorite strewn field. The fall zone has the form of an elongated ellipse. The stone moved across the sky from Southeast towards the Northwest..

Near Winslow, Arizona, Meteor Crater in the snow.

THE CONTROVERSY OVER THE ORIGIN OF METEOR CRATER

In 1891, the American geologist G. K. Gilbert saw a possible connection between a rimmed crater in northern Arizona, about 1,300 m across and 175 m deep, with iron meteorites scattered over the surrounding plains. He imagined that a gigantic body of iron had excavated the crater and buried itself in the floor. In the field, however, he failed to detect a magnetic anomaly or to find any clear evidence of an impact. So, despite the absence of volcanic rocks at the crater, Gilbert concluded that it had been formed by a deep-seated steam explosion, which he linked to the volcanism of the nearby San Francisco Peaks.

In 1902, the mining entrepreneur, Daniel Moreau Barringer, who strongly believed in the existence of a huge mass of buried iron, bought the mining rights to the crater. He and his partner discovered an impressive number of proofs of impact which are accepted today, but at the time they could not convince the geological community to override the authoritative opinion of G. K. Gilbert.

The controversy lasted for decades, while Barringer sank shafts and drilled holes without finding any trace of a buried meteorite except sludge rich in nickel-iron at a depth of 36 m. For a while, isolated craters seemed to be so mysterious that astronomers attributed them to a geological process (volcanism) and geologists to an astronomical process (impact). However, calculations began to show that large meteorites passing through the atmosphere at cosmic velocities will strike the Earth with explosive force: vaporising the projectiles and excavating craters such as those on the Moon and the one in Arizona. Beginning in 1928, discoveries of additional craters associated with meteorites convinced many people that Barringer was right about the impact origin of "Meteor Crater."

View of the inside of Meteor Crater with, at the centre, what is left of the exploration equipment used by Barringer.

Théodore Monod and the Chinguetti meteorite

From 1932 to 1998, the late Théodore Monod searched indefatigably for the giant meteorite of Chinguetti. It was described as forming a cliff 40 m high—very different in scale from even the largest of known meteorites. The meteorite supposedly was discovered in 1916 by Lieutenant Gaston Ripert, but no one has succeeded in finding it again. The field area, a vast region partly covered with dunes south of Chinguetti in Mauritania, does not make the task easy. Does the giant meteorite really exist? Professor Monod became convinced of an error on the part of G. Ripert, who, in the twilight, must have confused a sandstone cliff (the hill of Aouinet N'Cher) polished by wind erosion, with a gigantic mass of metal. However, at the top of the cliff the lieutenant did find a small 4.5 kg specimen that really is a meteorite (a mesosiderite). making it hard to doubt his good faith or his talents as an observer.

Théodore Monod on Aouinet N'Cher hill, in Mauritania, in 1993.

Debates continued into the 1950s before a consensus formed that many craters and "cryptoexplosion" structures with no associated meteorites also are formed by impacting meteorites.

THE COLLECTION OF METEORITES IN HOT AND COLD DESERTS

The majority of the meteorites that fall on the Earth every day are lost in the oceans

(~70% of the Earth's surface). Since meteorites fall at random, nobody imagined that some regions would be especially favourable for meteorite hunting until the discovery, about thirty years ago, in the semi-arid plains of Roosevelt County (New Mexico), of numerous meteorites exposed on the floors of so-called "blowouts" – shallow depressions swept bare of soil by wind erosion. Since then, searches in other windswept regions of hot deserts have met with remarkable success, notably parts of the Sahara desert in Algeria and Libya. To date, the most productive hot desert in the world is the Nullarbor plain of south-western Australia, where nearly 2,000 were recovered within the past ten years. As a single fall can yield multiple

fragments, it is difficult to estimate the exact number of falls this number represents. (All fragments of a fall share the name of one meteorite.) Most of the meteorites are ordinary chondrites, but some very rare types also have been found. In 1991 the Nullarbor plain yielded a meteorite from the Moon.

In 1969, Japanese glaciologists brought nine small fragments of rock found on a blue ice field in Antarctica back to Japan, where analyses showed them to be pieces of four entirely distinct classes of meteorites

Left: fragment of a meteorite found in Oman, along with 19 others that lay within a radius of 2 m. With its black fusion crust, the meteorite is distinguished particularly well from the light-coloured limestone beneath it.

Below: two searchers (including the author lying on the ice) have just discovered a meteorite. This record gives an idea of the small size typical of specimens collected in the Antarctic.

that had fallen to the ice at different times. This news caused a sensation! Here was the first known example anywhere on the Earth of a concentration of meteorites. Clearly, the motions of the ice sheet, which slides little by little towards the sea, had brought together the frozen-in fragments of different meteorite falls. Scientists assumed that

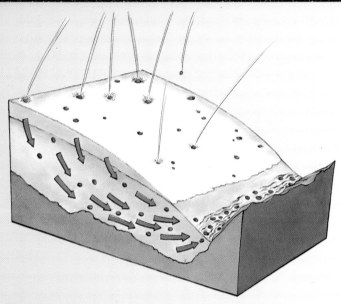

The concentration mechanism of Antarctic meteorites

As a general rule, meteorites that fall on the ice are progressively frozen into it and carried towards the shore, from which they float away in icebergs and finally fall to the bottom of the sea. However, if a volume of ice is trapped behind a mountain range, it will be eroded, little by little, by the violent winds that course down from the polar plateau and run along the bases of the cliffs, slowly exposing the meteorites at the surface. Since 1973, such stretches of stagnant ice have supplied incredibly rich concentrations. More than 17,000 fragments have been collected by diverse expeditions led by the Japanese, the Americans and, more recently, the Europeans.

similar concentrations should be found on other ice patches and so annual collecting expeditions began very soon. Among the most extraordinary Antarctic finds are the first lunar meteorites (13, to date) recognised on Earth, six coming from Mars, and two meteorites that fell about 2 million years ago, which makes them by far the oldest ever recovered. For comparison, the majority of finds date from around 100,00 years in the Antarctic, and less than 50,000 years for the other continents. The Antarctic thus constitutes a precious treasure chest for finding samples of asteroids, the Moon and Mars, while we wait for future missions into space...

Key words: **asteroid • basalt • fireball • chondrite • crater • crust of the Earth • dust • fall • find • fusion crust • impact • iron • meteor • magnetic grain • magnetite • meteorite • pallasite • pyrite(s) • shock wave**

Impact

craters

The smallest telescope allows us to observe a multitude of craters on the Moon. They are the signs of meteorite bombardment of the surface for billions of years. The phenomenon is very general in the solar system, but the example of the Earth shows that erosion and plate tectonics can progressively erase these cosmic scars.

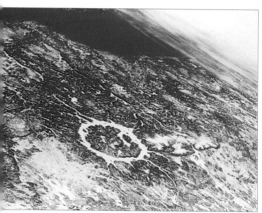

Above: terrestrial crater at Manicouagan, Canada, with a diameter of 100 km.

Left page: view of Mercury, with highly cratered surface. In the equatorial zone of the planet, we see the crater Kuiper, the brightest one on the planet.

Since the beginning of the space era, all the main planets of the solar system (except Pluto), their satellites, a comet and several asteroids (e.g. Gaspra and Ida), have been studied and photographed up close. Astronauts have walked on the Moon. Images taken by satellite have encouraged geological study of the Earth, and the number of impact craters (due to collision with a meteorite) identified on our planet has gone from about ten in the sixties to more than 150 today.

A UNIVERSAL BOMBARDMENT

The comparison of data gathered for the solid bodies of the solar system shows that practically all of them display craters on their surfaces, but that some are much more cratered than others.

Why do we observe only about 150 craters on Earth and a few hundred on Venus, while Mercury, Mars and the Moon, though smaller, have thousands of them? It is lunar exploration that has allowed us to answer this question. On the Moon, there is no erosion, no soil movement: a crater, once excavated, is eternal, unless it is later destroyed by a bigger crater or covered by lava flows. Thanks to the dating of lunar samples returned by the Apollo missions, we have been able to determine the meteorite flux over the course of time: it was very high in the first five hundred million years of the solar system, but much weaker since. The Earth and the Moon, close neighbours, have been bombarded in similar fashion. But the vast majority of terrestrial craters have disappeared because the surface of our planet continually renews itself due to the effect of erosion and plate tectonics. On the

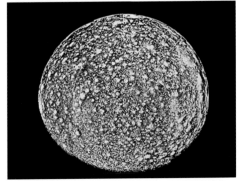

Moon, on the other hand, certain regions appear saturated with craters because they have been unchanged for more than four million years. Other bodies, like Venus, are relatively less cratered, and thus have an intermediate surface age, indicating that they must be resurfaced.

COSMIC CHRONOMETERS

How can we determine the age of a surface as a function of the density of craters observed on it? The problem is a tough one, because we do not have available samples from the bodies concerned, except in the case of the Moon, which would allow us to know, for each, the relationship between age and crater density. We are reduced to using models established in the following fashion: we know that craters result from the impact of asteroids and comets; we calculate what the flux should be which reaches the different planets and their satellites, taking into account their position in the solar system, the respective force of gravitational attraction, from that which has been measured for the Moon, and from what we know of the distribution of asteroids and comets in the solar system. All we need to do then is to count the craters we see on any body in the solar system to be able to assign a surface age, which depends

Above, from left to right and top to bottom, views of cratered planets and satellites in the solar system: the Moon, Mars; Callisto and Dione.

on the model used. This is how we say, for example, that some volcanoes on Mars have been extinct for more than a billion years, with an uncertainty of at least 500 million years. However, it is enough if we are mistaken in the supposed distribution of comets and asteroids that all the age estimates we give will be wrong! To verify the validity of the models, it would be extremely useful for space missions to return samples from Mars and Mercury to Earth, for subsequent dating in the laboratory.

From impact to crater

From its velocity, a meteorite has a considerable kinetic energy on arriving at the surface. This is transformed partly into thermal energy, but especially into mechanical energy. The thermal energy more or less vaporises the meteorite and melts 5 to 10% of the debris derived from the impact. However, it is the mechanical energy which gives rise to the most

significant effects. Colossal pressures, of an intensity reaching up to several million times atmospheric pressure, are exerted in the impact zone. Then the overpressure propagates into the ground like a wave which is attenuated only very slowly. On its passage, certain minerals are vitrified and others recrystallise (the rocks so transformed are termed impactites); the underlying rock fractures forming particular structures, shatter cones. However, the formidable thrust that it exerts on the deep basement runs slap into the resistance of all the underlying planetary material and thus does not produce spectacular effects. On the other hand, once the wave has passed, the zone which was compressed relaxes brutally and, as this

Cosmic bombardment over the course of time

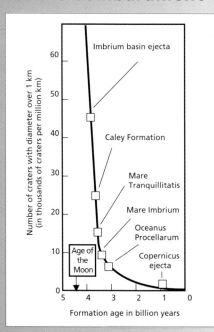

Cratering curve showing the number of craters per unit area as a function of the age of these surfaces (lunar data).

The lunar samples brought back by the Apollo missions have been dated in the lab by the usual techniques based on radioactivity. For the six regions of the Moon which were visited by astronauts, we therefore know both the number of craters per million square kilometres (determined from photographs taken in orbit) and the ages of these terrains. We have thus been able to establish the relationship between the age and the extent of cratering, which has two extremely interesting consequences: 1) it is now sufficient to determine (by telescope, for example) the extent of cratering of any lunar terrain to know its age. 2) we have been able to evaluate the flux of projectiles (that is to say the number of meteorites fallen per million square kilometres per year) on the Moon since 4 billion years ago: this flux, large at first, decreased strongly and, for the last 3.5 billion years, has been very low.

Impact Rocks

Impact breccias are are made up of fragments of the target rocks, sometimes with splashes of impact melt, sometimes with more or less shocked mineral inclusions. With greater impact energy, we go from breccias with a matrix of fine mineral fragments, to glassy-matrix breccias, fragment-laden impact melt rocks and impact melt rocks superficially resembling normal igneous rocks. Most impact rocks are easy to link to their source material from their chemical, mineralogical and isotopic characteristics.

meteorites arrive at an oblique angle? Because they result from the intersection of a planet and a wave front, both spherical, and the intersection of two spheres is always a circle. Laboratory simulations and calculations allow us to estimate the energies brought into play. For Meteor crater (Arizona) we find, for example, an energy equivalent to that released by the explosion of 1,700 kilotons of TNT, or 133 Hiroshima bombs (diameter 1.3 km). For the Chicxulub crater (Mexico), the largest meteorite crater known on the Earth, it is even bigger: at least 5 billion tons of TNT, or more than 450 million Hiroshima bombs!

CRATER MORPHOLOGY

time nothing prevents its upward movement, this leads to ejection of material and the formation of a crater. Why are almost all craters circular, when many

Small craters, whose diameters do not exceed a few kilometres, are always bowl-shaped, with a depth equal to about a tenth of their diameter. In fact, they are almost

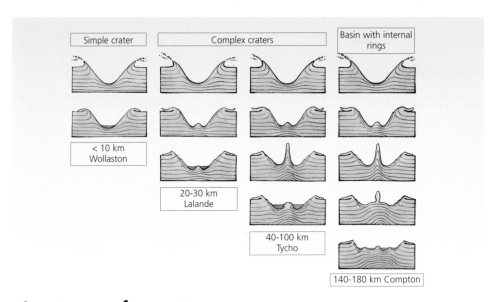

Simple crater | Complex craters | Basin with internal rings

< 10 km Wollaston

20-30 km Lalande

40-100 km Tycho

140-180 km Compton

Anatomy of a crater
These diagrams show the formation of complex craters and basins with internal rings. The numbers indicate the diameters of the corresponding lunar craters.

twice as deep, but partially refilled by ejection debris (ejecta) which have fallen back into the interior. They are also found surrounded by a blanket of ejecta. On planets or satellites without atmosphere, the debris can be projected very far and sometimes traces radial structures around the crater: these rays are particularly well seen around the youngest lunar craters, like Tycho or Giordano Bruno. Beyond a certain diameter (3 km on Earth, 10 km on the Moon), the morphology of craters becomes very different: their bottom tends to flatten and shows a central peak or even, in the biggest, a ring. We then speak of complex craters. Peak and ring develop because, after the impact, the bottom of the impact is projected upwards, like water in a puddle when it is bombarded by big raindrops. After the rebound, the walls of the crater collapse, thus forming a series of terraces: these are very visible, for example, inside the crater Copernicus, on the Moon. The phenomena responsible for the morphology of craters depend strictly on the gravity of

A simple crater 4 km in diameter on the Moon, in Mare Serenitatis (equivalent to 5,000 Hiroshima bombs).

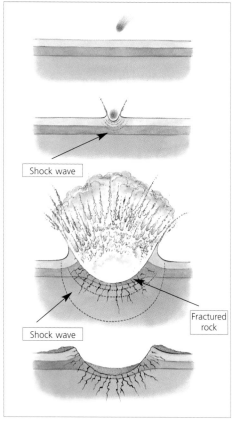

Shock wave

Shock wave

Fractured rock

Diagram of the formation of a simple crater.

On the Moon, the complex crater Theophilius, 100 km in diameter, with central peak.

Mosaic showing a view of the lunar surface and its numerous little craters and regolith.

Lunar glass sphere seen in the scanning electron microscope, with a microcrater 0.1 mm in diameter.

the body on which the impacts take place: this is how we explain the differences between terrestrial craters and lunar craters.

Up to now, we have only talked about the effects of big or very big meteorites. But micrometeorites, infinitely more abundant, have an entirely analogous behaviour, on a different scale, when they hit the surfaces of bodies without atmosphere: they excavate microcraters on them. The latter overlap and obliterate one another in their millions, over the course of time. The ground is thus progressively completely pulverised, and overturned for a depth of several metres, as if tilled by a gigantic plough. We then obtain

The Steinheim crater, in Bavaria, 3,500 km in diameter, including a central peak which is very visible because the rock which forms it is bright in the infra-red.

what we call a regolith, of which the lunar soil is a perfect example. Certain meteorite breccias consist of compacted regolith.

THE LUNAR MARIA

The side of the Moon turned towards the Earth is strewn with great dark stretches, often circular, visible even with the naked eye. These are the famous lunar seas, following a terminology that goes back three hundred years, to a time when their appearance was explained by the presence of water. In fact, these are vast discharges of basaltic lava, which fill ancient impact basins. Should we conclude that there is a direct relationship between impact and volcanism? Observations obtainable by telescope, within the capability of every amateur astronomer, show there is not: we note, for example, that four craters (Sinus Iridium, Plato, Cassini and Archimedes) overlap the edge of Mare Imbrium or perforate its interior, while being themselves covered or filled with lava. These four structures are thus younger than the enormous impact which created Mare Imbrium, but older than the end of the volcanic episode responsible for the

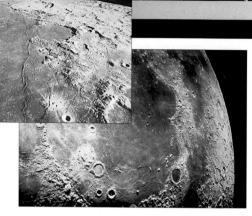

Left: contrast between lunar mare and highland, with the eastern border of the little-cratered Mare Serenetatis (age: 3.5 billion years) and the highlands.

Right: Mare Imbrium shows the non-contemporaneity between the giant impact (equivalent to 100 billion Hiroshima bombs) and its filling by mare basalt.

discharges of lava. Dating of samples brought back by the Apollo 14 and Apollo 15 missions confirm that the formation of Mare Imbrium by the impact of an enormous meteorite preceded the end of its filling by lavas by 500 million years. There was thus no direct relation between the two phenomena: there are geological reasons, notably the small initial thickness of the crust beneath the craters, which facilitated the arrival and discharge of lava.

A GREAT DIVERSITY

When a meteorite falls on a surface containing frozen water, as is the case in different regions of Mars, the ejecta form a blanket around the crater which is clearly

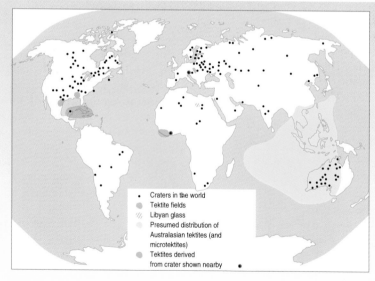

- • Craters in the world
- ● Tektite fields
- ⁄⁄⁄ Libyan glass
- Presumed distribution of Australasian tektites (and microtektites)
- ● Tektites derived from crater shown nearby •

Above: an australite button, right, has been photographed beside a small meteorite, left. Traversing the atmosphere at great speed during their fall, they have acquired more or less the same shape, after melting partially (ablation).
Left: map of the world showing crater distribution and tektite fields.

Tektites

Tektites are glasses produced at very high temperature (more than 2000 °C) during an impact producing a crater with a diameter greater than 10 km. Ejected to great altitudes (in the absence of atmosphere), some spheres have retained enough energy to be spread over hundreds, even thousands of kilometres (australites).

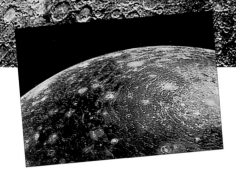

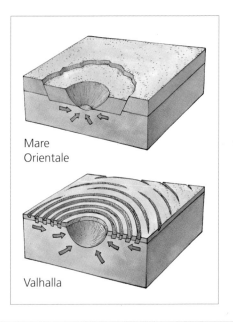

Below: two schematics for the formation of basins with internal rings (peaks and internal rings are not shown). The first basin corresponds approximately to Mare Orientale, on the Moon, and the second to the Valhalla basin on Callisto (photo at left), whose original cavity, with a diameter of 300 km, is surrounded by about 15 external rings (equivalent to two billion Hiroshima bombs).

delimited instead of gradually thinning with distance from the impact site. We get a similar pattern by throwing a stone in a puddle of mud. Indeed, the ejecta, mixtures of crushed rock and melted ice, form a sort of thick mud which, on flowing out around the crater, gives it its particular morphology. Multi-ring basins are even odder, with central cavities (sometimes with a peak or peak ring) surrounded by mountainous concentric rings. Thus Mare Orientale, on the Moon, consists of a cavity 600 km in diameter, with two internal rings due to rebounds and slumps of its central peak: its external boundary is a gigantic fault which formed 150 km beyond the edge of the crater. The record for complexity is held by the Valhalla basin, on Callisto, one of the four main satellites of Jupiter. In this one there is a central cavity 350 km in diameter,

Mare
Orientale

Valhalla

A meteorite in a meteorite

The meteorite which fell on May 30th 1866 at St. Mesmin, in France, is remarkable from several points of view. The consolidated dust and fragments which make it up were situated on the very surface of the LL asteroid. This had a diameter of about 250 km. Like all bodies devoid of atmosphere and exposed to space, it was immersed in the solar wind, which irradiated crystals for a depth of about a ten thousandth of a millimetre. Since its formation, the surface of the asteroid was hammered by the impacts of objects of all sizes. The most massive projectiles extracted deep material which smaller ones could later break, erode and round, producing a dusty regolith which ends up here in consolidated form. Despite the considerable lowering in the flux of projectiles, the asteroid was struck, 1.3 billion years ago, by a chondrite, a chip of which has got to us. Its slightly different chemical nature from the material in which it is enclosed means that we spot it immediately with the eye: it is dark and speckled with bright metallic spots – it is an H chondrite. Radioactive chronometers allow us to date the stages in this history.

The Tunguska explosion

On the morning of June 30th, 1908, S.B. Semenov was hooping a barrel in front of his house, in the little town of Vanavara, on the banks of the Tunguska, in central Siberia. Suddenly, the northern half of the sky, in front of him, blazed. He felt a strong heat, then a deafening explosion resounded, he was thrown to the ground and lost consciousness. A celestial object, 70 km further north, had just exploded above the Siberian forest. The first scientific expedition did not arrive on the spot until nineteen years later. It found the forest devastated over nearly 2,000 km², but no meteorite. The numerous expeditions which followed one another to the site did not have better luck. The entire fireball was probably dispersed in the atmosphere. The energy of the explosion is estimated as a dozen megatons of TNT, but we have not been able to establish with certainty the nature of the object which exploded: comet fragment, ordinary chondrite, carbonaceous chondrite... If we assume that it moved at 15 km/s, a typical velocity for a meteorite, its mass must be about 500,000 tons. If it was an ordinary chondrite, its diameter was of the order of 60 m.

surrounded by more than 15 concentric faults which trace rings visible up to 1,500 km from the edge of the crater.

TOOLS FOR GEOLOGICAL INVESTIGATION

The solid planets and satellites have a superficial cold and rigid envelope, the lithosphere, overlying a hotter, less rigid layer, the asthenosphere. Given a meteorite impact, if the resulting crater does not reach the asthenosphere, the effects of the shock are only manifested inside the cavity (peak or central ring). But if the crater is sufficiently deep to pierce the lithosphere and penetrate into the asthenosphere, the consequences are more dramatic. The solid but viscous material will slowly come and fill the hole, which will exert a drag on the base of the lithosphere and will cause the concentric faults observed outside the crater. The study of craters thus allows us to study the internal structure of planets.

Meteorite impacts represent geological processes and tools of major importance in the solar system. The only great fall observed so far, that of the fragments of the comet Shoemaker-Levy 9 on Jupiter in July, 1994, happened on a gaseous planet. Thus it did not cause any craters. Perhaps tomorrow another fall, on a solid planet (preferably not the Earth!), will allow us to better understand the mechanisms of formation of impact craters.

Key words: **fireball • breccia • chondrite • comet • crater • cratering • crystallisation • ejecta • glass • impact • impactite • impact basin • isotope • meteorite • micrometeorite • plate tectonics • radioactive chronometer • regolith • shatter cone • shock wave • tektite**

Cretaceous Park

Sixty five million years ago, numerous living species, in particular the dinosaurs, abruptly disappeared. Several lines of evidence today support the hypothesis that this ecological upheaval was provoked by the fall of a celestial body several kilometres in diameter on Earth.

THE PALAEONTOLOGICAL FACTS

As early as 1812, Georges Cuvier, the founder of vertebrate palaeontology, deduced from observations made on the chalk of the Paris basin that major faunal modifications occurred at a time now regarded as the end of the Mesozoic era. Since Cuvier, geology and palaeontology have made immense strides. We now know that the passage from the Mesozoic to the Tertiary, dated since 1992 at exactly 65.0 million years, was extremely abrupt. This sharp transition is what we call the Cretaceous-Tertiary (K-T) boundary. It left its mark particularly visible in marine sediments: more than 80% of the species making up plankton (the little single-celled vegetables and animals living in the surface water of the oceans) of the Cretaceous disappeared simultaneously, apparently along with some molluscs like ammonites, belemnites and rudists.

On the continents, the end of the Mesozoic was also marked by spectacular upheavals, with the disappearance of all the animals bigger than 25 kg, in particular the dinosaurs. However, unlike planktonic species, the big animals have not left sufficiently numerous fossils that their evolution can be followed in a continuous fashion. Some think that all these animals showed signs of decline and disappeared well before the end of the Cretaceous; others, however, consider that their disappearance was brutal and without prior warning. The most recent studies seem to show the latter correct: the last fossil ammonites and dinosaur footprints were found in the last

Above: fossil ammonite (Acanthoceras cenomanense) and certain foraminifera (marine mico-organisms) which are found in Cretaceous sedimentary beds but no longer exist in the Tertiary.

Left page: Earth encounter with an asteroid of about 10 km size: painting by W.K. Hartmann.

Above: close-up of the K-T boundary at Hendaye (France) The fine bed of dark greenish clay is enriched in iridium and contains crystals of nickel-bearing magnetite and shocked quartz.

Right: general view of the cliff at the Stevens-Klint site (Denmark). The arrows indicate the fine bed of clay separating the Cretaceous from the Tertiary.

few metres or decimetres of the Mesozoic sediments, in beds which were deposited less than a few tens of thousands of years before the K-T boundary.

To explain the upheavals experienced at the end of the Mesozoic, numerous hypotheses have been proposed. We think particularly of the intense volcanic activity which reigned at the end of the Cretaceous and the general lowering of sea level (marine regression) observed at that time. Indeed, these two phenomena, whose existence is scientifically proven, are likely to have perturbed the conditions for life over the entire Earth in a major way and, therefore, to have been accompanied by the disappearance of certain species. However, these were long lasting phenomena: they could not then explain the sudden disappearance of the planktonic species.

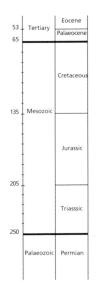

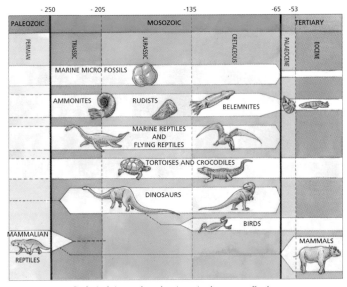

Geological time scale and major animal groups on Earth.

AN IMPORTANT BUT AMBIGUOUS TRACER: IRIDIUM

A considerable advance was made at the beginning of the eighties with the discovery on a global scale of iridium in marine sediments at the very end of the Cretaceous, very exactly at the level of the clay marking the sudden disappearance of the majority of the planktonic species. Iridium, like all the platinum group elements which are associated with it in mineral deposits, is practically non-existent on the surface of the Earth. On the other hand, it is more abundant in meteorites and the depths of our planet (in the mantle and core). Its appearance in the sediments of the K-T boundary, at the very time when numerous planktonic species disappeared, implies a common origin for these two phenomena.

Would that be the lowering of sea level? No, because that could not explain the briefness of the event nor the presence of iridium over the entire planet. Some authors have tried to explain the presence of iridium by a biological process: blue algae are capable of concentrating iridium carried by the constant rain of meteoritic dust. This hypothesis, formulated to explain the iridium observed at the Frasnian-Famenian boundary in Australia (360 million years ago), cannot in any way explain what we see at the K-T boundary. On the one hand, no anomaly on a global scale has been observed for the Frasnian-Famenian boundary: the presumed biological activity is thus only a local phenomenon. On the other hand, the total amount of iridium at th K-T boundary (500.000 tons) represents the flux of micrometeorites integrated over more than a hundred million years. Such a quantity could only be explained by an exceptional supply of extraterrestrial material or an

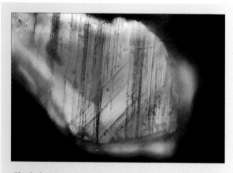

Shocked quartz seen in the optical microscope with crossed polarised light. The criss-crossing lamellar structures show the presence of quartz made amorphous by the passage of a shock wave. This crystal was found at the Bidart site (France).

Shocked quartz

The collision between the Earth and an asteroid is an event of extreme violence. A considerable energy is liberated in less than a second and deformations are produced by the very great pressures induced during the impact, by the propagation of a shock wave into the Earth's crust. Quartz is a mineral which retains particularly well the memory of this exceptional mechanical stress in its crystal lattice. Shocked quartz grains contain very fine lamellae of amorphous (that is to say non-crystalline) silica, which is very rare in nature. Laboratory experiments have shown that we can only produce such lamellae by a very rapid pressure change (to a pressure 10,000 times atmospheric pressure). The only natural phenomenon capable of inducing such shock waves is a collision between the Earth and a great meteorite. Shocked quartz therefore represents an important index mineral in geology, allowing us to identify ancient impact craters which, because of erosion, are generally difficult to detect.

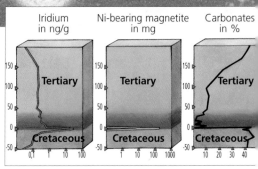

enormous eruption of lava from the deep interior. However, chemical and isotopic analyses cannot distinguish between these two potential causes. To remove the ambiguity, we have had to use other markers, minerals.

THE MINERALS TALK

The sedimentary beds at the K-T boundary are not characterised only by an abnormally high concentration of iridium. They also contain special mineral grains, of which the two most important types are shocked minerals and magnetic minerals.

Shocked grains, essentially quartz, have sizes between a few tens and several hundred microns. In cross-polarised light, they show unusual criss-crossing lamellar structures. For a long time controversial, the origin of these structures is now well established: only a shock wave, that is to say a pressure wave moving quicker than sound, can cause such deformations in the

Stratigraphic distribution of iridium, Ni-bearing magnetite and carbonates across the K-T boundary at the d'El Kof site (Tunisia). We note that the distribution of magnetite is extremely restricted. This shows that the supply of cosmic material happened almost instantaneously, coinciding with the disappearance of plankton. The spread of the iridium anaomaly is due to geochemical effects (diffusion) and does not indicate the real length of the event. Goethite level in red.

crystal lattice of quartz. In fact, volcanic eruptions are incapable of producing such effects. The presence of shocked quartz in K-T boundary sediments therefore demonstrates mechanical effects induced by impact, at this time, of one or several big extraterrestrial objects – comets or asteroids.

Nickeliferous magnetite from the Caravaca site (Spain) seen in the scanning electron microscope. The ocatahedral form is common in K-T boundary sediments but dendritic forms are also found.

Nickel-bearing magnetite

Nickel-bearing magnetite is a mineral whose formation requires melting at more than 1,300 °C of a material rich in nickel in an atmosphere rich in oxygen. While these two conditions are never satisfied in terrestrial magmas, they always are when a meteorite, rich in nickel, passes through the atmosphere: the external surface of the meteorite heats up and experiences significant erosion (ablation); fine droplets of molten material are torn off by aerodynamic friction. The Ni-bearing magnetite crystallises in these droplets, which become oxidised on contact with the atmosphere. The composition of the magnetite depends on the oxygen pressure, and therefore on the altitude at which the oxidation takes place. The composition of the K-T boundary magnetite corresponds to extremely oxidising conditions.

We can add another indicator to this. The K-T boundary sediments contain a unique magnetic mineral with no equivalent in terrestrial rocks. This is present in the form of little 1 to 20 micron crystals of nickel-bearing iron oxide. This mineral is the tracer of extraterrestrial material which has oxidised in the atmosphere before reaching the ground. Its presence in sediment at the end of the Cretaceous demonstrates the meteoritic origin of the iridium with which it is associated. Its stratigraphic distribution shows, moreover, that this supply of meteoritic material happened in a very short expanse of time, less than a few tens of years.

A COSMIC COLLISION

Thus all the observations converge remarkably and show that the end of the Mesozoic was marked by the collision of a celestial body about 10 km in diameter with the Earth. The formidable energy liberated by the impact vaporised most of the projectile and a fraction of the target, and dispersed the vapour and accompanying debris over the whole planet. This material, now much modified, is no longer recognisable except from the presence of iridium and the most resistant mineral, Ni-bearing magnetite. The ejected target

Left: fall of a meteorite on Earth by S. Numazawa.

Below: scenario showing the consequences of the fall on the Yucatan peninsula: earthquakes, giant waves, transport of ejecta debris, forest fires on a global scale, dust cloud in the atmosphere attenuating sunlight, thus diminishing photosynthesis and cooling the planet, acid rain over the whole globe and greenhouse effect caused by gas evolved from acid fallout.

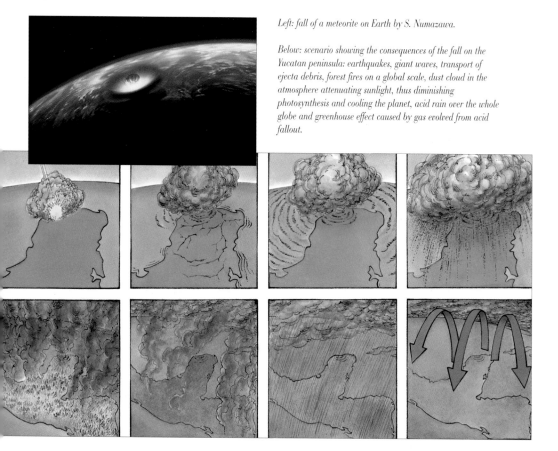

The Chicxulub crater

The great abundance of shocked quartz crystals and the presence of impact glasses in Central and North America led us to believe, for a long time now, that the impact at the end of the Cretaceous happened in this region of the globe. Several sites were considered at first. The Manson crater, in Iowa, in the United States, provoked research the longest, even though it was a bit small for the presumed size of the fireball, but the precise determination of its age (72 million years) led finally to its elimination. Attention was then turned to the Columbia River Basin and the vicinity of Cuba. Finally, it was in Mexico, in the Yucatan, that the crater was tracked down, thanks to petroleum exploration and, especially, measurement of gravitational anomalies: the latter reveal several circular structures, of which the largest is nearly 200 km in diameter. Centred on the village of Chicxulub, the crater is buried beneath several kilometres of Tertiary sediment. The analysis of melt rocks sampled by drilling has allowed the crater to be dated at 65.0 million years. This is exactly the age of tektites found at the K-T boundary at sites in Haiti and Mexico. This result has been confirmed by the study of shocked zircon (zirconium silicate).

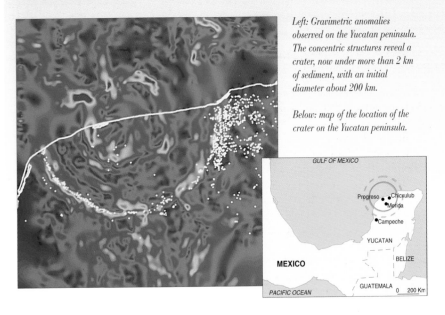

Left: Gravimetric anomalies observed on the Yucatan peninsula. The concentric structures reveal a crater, now under more than 2 km of sediment, with an initial diameter about 200 km.

Below: map of the location of the crater on the Yucatan peninsula.

material carrying the evidence of impact, shocked minerals and impact glasses (or tektites), however, has a much more limited distribution. The great abundance of shocked quartz crystals and tektites observed around the Gulf of Mexico suggest that the collision happened in this region.

An enormous circular structure more than 200 km in diameter, revealed by anomalies in the gravity field, in the north of the Yucatan peninsula, around the village of Chicxulub, appears to be the impact crater, now buried beneath hundreds of metres of Tertiary sediment.

Reconstruction of Tarbosaurus, Upper Cretaceous, the remains of which were found in the Gobi desert (Mongolia). The tarbosaur is very close anatomically to the tyrannosaur found in North America.

The effects of such a collision can only have been disastrous and proportional to the energy liberated, equivalent to that liberated by more than 500 million Hiroshima bombs. Some immediate effects are still visible today. Near the crater (within a radius of a few hundred kilometres), we see a layer of debris and fractured rock (breccias) several metres to several tens of metres thick. Further afield, over thousands of kilometres, we find traces of the earthquake and giant wave which ravaged the shores of the Gulf of Mexico and neighbouring regions. Finally, on a planetary scale, we note the presence of soot, which can be explained by the combustion of more than half the vegetation which existed at the end of the Cretaceous.

MURDEROUS CLIMATIC UPHEAVALS

What mechanisms caused the disappearance of numerous species following the collision? The problem is far from being understood. In particular, it is hard to explain at the present time why some species disappeared and other similar ones survived.

The shower of debris ejected by the impact must have caused a blast of heat over the entire Earth surface for a few hours. Perhaps it was this "heat wave" which set off the widespread burning of forests and caused the formation of soot. It is also possible that the enormous quantity of dust injected into the atmosphere by the impact, by absorbing the Sun's light, caused global cooling of the Earth for several years and slowed photosynthesis on the planet.

From the palaeontologists

Skeleton of Tarbosaurus bataar from the Upper Cretaceous.

Was the extinction of the dinosaurs and other species at the end of the Cretaceous gradual or abrupt? Were there a series of environmental changes during the late Cretaceous affecting biodiversity, to which the Chicxulub impact was a minor addition? Does the impact event really come exactly at the end of the Cretaceous? Although the widespread ejecta layer of melt spherules has exactly the same age, 65.0 million years, as the impact melt in the Chicxulub crater, near the crater the spherules were covered by additional sediments before the deposition of material with Tertiary fossils. These additional Cretaceous sediments have been interpreted by some as backwash from the land due to the giant impact wave (megatsunami). At Bass River, New Jersey, the picture is much clearer. The spherule layer, up to 100 cm thick near the crater, is here only 6 cm thick. The spherules make imprints in richly fossiliferous Cretaceous clay and are overlain by clay with the basal Tertiary fossil assemblage. A complete Cretaceous-Tertiary boundary transition is present, and all except three species of the Cretaceous planktonic foraminifera become extinct at the spherule layer. It is thus clear that the Chicxulub event is directly linked to the mass extinction of marine biota. How did such a sudden catastrophe exterminate dinosaurs and not crocodiles, rudists and not oysters, flying reptiles and not birds? Crocodiles are low-budget energy animals (ectotherms), do not need as much food as dinosaurs to maintain their metabolisms, and are capable of "shutting down" for extensive periods of time when environmental conditions are not favourable (cold). Crocodiles, unlike dinosaurs, could hibernate through the global winter. Rudists were reef ecosystem elements, and reefs are notoriously sensitive to environmental conditions. They are the first to go in any biotic crisis. Oysters, on the other hand, are simple benthic filter feeders. It is a mystery as to why the birds survived at all. Birds are very delicate organisms, both anatomically and ecologically, and are endotherms. The biggest factor in their favour (for survival) is that they are small — and in an environmental catastrophe, "big" is "bad" news. Thus, it appears clear that a sudden environmental perturbation by a major impact caused the terminal Cretaceous extinctions.

we see unexplained selective effects. At the bottom of the sea, on the other hand, where the animals make do with organic debris, the catastrophe passed almost unperceived.

We can imagine other major effects, which have not been entirely documented: acid rains coming either from sulphur dioxide given off by the heating and breakdown of rocks in the impact crater, or the considerable quantities of nitrogen oxides formed in the atmosphere after the catastrophe.

FROM REPTILES TO MAMMALS

The list of possible consequences due to the impact suffered by the Earth at the K-T boundary is so appalling that it makes us ask ourselves how life could have gone on after such a catastrophe. This is a problem to which we can give only partial answers: the details of the cataclysm are unknown for now and perhaps for ever. Anyhow, various observations lead us to believe that certain regions of the globe, in high latitudes, for example, could have been partly spared. It is probably from these natural reserves that life on our planet was able to replenish itself. Several hundred thousand years were necessary for a new state of equilibrium to establish itself, but biological activity never became again what it was at the end of the Mesozoic era. Irreversible changes were produced: reptiles no longer dominated the planet, the way is clear for mammals.

Why did the catastrophe responsible for the extinction of the ammonites (at the top) spare their close cousins the nautiloids (above)? The very special mode of reproduction of the latter (eggs laid at very great depth which take more than a year to hatch) may well have been the reason for their survival at the moment when the planktonic fauna and young ammonites were killed. Only a detailed study of the ecological characteristics of the diverse Jurassic species would allow us to understand the survival of some and the extinction of others...

This prolonged "winter" would have deprived herbivorous animals of food, especially the big eaters. Hence the disappearance of the dinosaurs. In the oceans, the prolonged absence of light would have prevented the development of vegetable plankton. However, here again,

Key words: **ablation • amorphous • asteroid • biodiversity • biosphere • breccia • comet • core • crater • cross-polarised light • crust of the Earth • crystal lattice • crystallisation • era • extinction • fireball • glass • gravitational anomaly • impact • iridium • K-T boundary • lamellar structure • magma • mantle • magnetic minerals • Ni-bearing magnetite • shocked quartz • silica • tektite • zircon**

■ A piece of the Allende meteorite (about 300 g and 10 cm along the base). Thousands of stones fell on February 8th 1969 in an elliptical strewn field of over 150 km^2 around Pueblito de Allende in Mexico. They are covered with a dull black crust (visible on the right of the photo). The broken surface reveals numerous white inclusions, often with irregular outline, which are visible against the dark grey background. We also see, but less easily, numerous little grey spheres: these are "chondrules" which make up much of this meteorite. The white inclusions have a particularly interesting and unique composition. Discovered the year before in a meteorite of the same type as Allende, their occurrence in this abundant material allowed us to undertake an in-depth study using the sophisticated analytical methods very recently perfected for the return of lunar samples. Year of the conquest of the Moon, 1969 was also a pivotal year for the study of meteorites.

ɔn Earth

Despite the diversity of rocks on Earth, none resemble the meteorites most often seen to fall: the chondrites. They have chemical compositions and physical characteristics derived directly from the nebula from which the solar system formed. In addition to chondrites, there are meteorites described as "differentiated" which come from objects (asteroids, planets) which, like the Earth, were melted.

■ Right: the Bouvante stone is a pale compact rock covered with a thin shiny black skin. This fusion crust is due to the heating of the rock by friction during its passage through the atmosphere. This rock does not contain any chondrules. It melted then crystallised like a terrestrial basalt.

9 cm

23 cm

■ Left: the Staunton iron. This slab of polished metal – iron-nickel alloy – containing two nodules of iron sulphide is a piece of an iron meteorite. This type of meteorite, which occurs as a dense irregular blackish mass (reddish when oxidised) does not resemble any terrestrial rock, but is easily confused with products of the metallurgical industry.

HOW TO RECOGNISE A METEORITE?

Many objects can be erroneously taken as meteorites. Two main characteristics distinguish most meteorites: they are covered with fusion crust (related to the passage through the atmosphere), and contain metallic minerals, which makes them magnetic. But the identification can be more difficult; the fusion crust can be absent from a fragment broken on the ground, or can have more or less disappeared under the effect of terrestrial alteration. Apart from diverse products of human activity such as foundry wastes and shrapnel, we can find stones attracting a magnet which, even so, are not meteorites. A chemical test is needed, to reveal the presence of nickel in the metal, coupled with the absence of numerous other metals used industrially. We can also try to find "Widmanstätten patterns"

10,5 cm

■ Above: the Douar Mghila stone is covered with a continuous crust. In this picture, we see trails of melted material that formed near the central hump, corresponding to the forward face of the meteorite during passage through the atmosphere. We rarely find a stone so clearly oriented: often blocks split in the high atmosphere, change direction, and the crust forms simply a black coating less than 1 mm thick, glossy on the surface of achondrites like Bouvante, matt on the surface of chondrites. When a stone has spent a long time on the ground, the crust can be abraded, rusty or even have disappeared.

10,5 cm

■ Above: polished slice of a chondrite (Flandreau). Here the crust, which surrounds the stone, is rusty. The number of grains of metal and sulphide (here white and greyish) allows us to recognise it as a meteorite. The stone is moreover denser than a terrestrial rock, because of the presence of these metallic minerals.

MINERALOGY SPECIFIC TO CHONDRITES

Chondrites take their name from spherical structures unknown on Earth which can constitute up to 80% of their material: chondrules (from the Greek chondrion: a granule). Certain chondrites (mostly carbonaceous chondrites) can also contain refractory white inclusions, made up of minerals rich in calcium, aluminium and titanium which are the carriers of isotopic anomalies. Chondrules and inclusions are held together by a finely crystalline material which may be opaque in transmitted light: the matrix. It is believed that chondrules, inclusions and matrix were all formed in the solar nebula. These different features are shown here in polished thin section (0.03 mm thick) seen in the optical microscope in cross-polarised or reflected light.

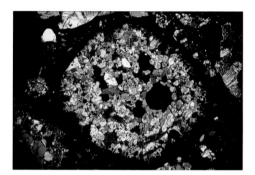 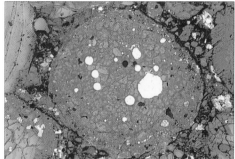

■ Above: the Chainpur meteorite (x 30). The centre of the photo shows a microgranular chondrule containing olivine and metallic globules. In polarised light (left), the little crystals of olivine (a silicate of iron and magnesium) form a multi-colour mosaic, while the metallic globules are opaque, or black. In reflected light (right), the silicates are grey, while the Fe-Ni globules are white. The chondrule rim is very fine-grained. In it we see silicates, metal , and pale yellow iron sulphide.

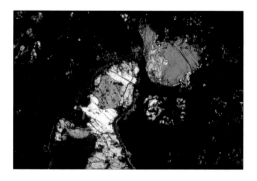

■ The Acfer meteorite (x 30). In cross-polarised light, we see the contrast between a fragment of olivine chondrule (purple and blue crystals) and a piece of a typical white inclusion, made up of melilite (a calcium alumino-silicate) appearing grey and white. The rim which bounds it (black at first and then frayed yellow) is specific to this type of inclusion. All the matrix which is black in the image is made up of very fine-grained material, in which the superposition of grains prevents the transmission of light.

WHAT IS DIFFERENTIATION?

Differentiation is the name we give to the process by which a mixture with an initially homogeneous composition separates into several phases with different chemical compositions. In the case of planets, an initial material of solar composition will be fractionated and separated into distinct layers: core (metallic), mantle (olivine-rich) and crust (basaltic).

Before differentiation

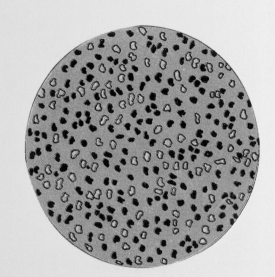

■ Right and top of page 57: schematic of the probable evolution of asteroids and planets.
– **A**: formation by accretion of a mixture of iron-magnesium silicates (green) with high melting temperature, minerals melting at lower temperature (yellow), and grains of metal and sulphide (black), in chondritic proportions. Chondrites are derived from such bodies, little or not at all changed.

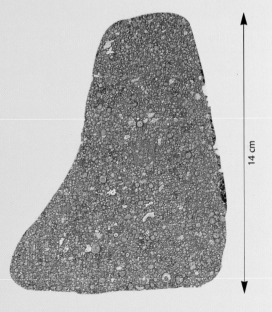

14 cm

■ Right: polished slice of Axtell, an undifferentiated meteorite of the same type of chondrite as Allende. We can see numerous refractory white inclusions with irregular outlines, chondrules are very abundant, and everything is cemented by fine crystalline dust. This meteorite is derived from an undifferentiated, and even little heated, asteroid.

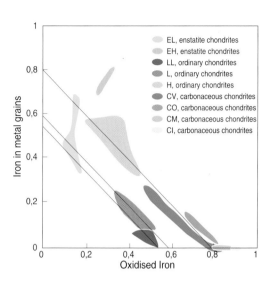

EL, enstatite chondrites
EH, enstatite chondrites
LL, ordinary chondrites
L, ordinary chondrites
H, ordinary chondrites
CV, carbonaceous chondrites
CO, carbonaceous chondrites
CM, carbonaceous chondrites
CI, carbonaceous chondrites

Iron in metal grains

Oxidised Iron

■ Right: iron in rocks can be in two main oxidation states: reduced (in metallic minerals) and oxidised (when combined with oxygen or sulphur, in silicates, oxides or sulphides). Here the amount of reduced iron is plotted against the quantity of oxidised iron. The diagram shows that the different classes of chondrites are chemically distinct.

■ Above: photo of the Forest Vale chondrite, taken in the scanning electron microscope. This photo lets us see chondrules in relief (size of the biggest one: 200 µm).

3,5 cm

■ Above: the Orgueil chondrite fell in 1864 in a little village near Montauban. Its mass (about 10 kilos) makes it an object of prime importance, given the rarity of meteorites of this type. Indeed carbonaceous chondrites of type CI are considered the most primitive as they have the composition closest to that of the sun. Paradoxically, they do not contain chondrules. We could argue that this is the result of alteration on the parent asteroid, but it is more likely that it is a question of material accreted long before or after the formation of chondrules, or in regions (further from the sun?) without chondrules.

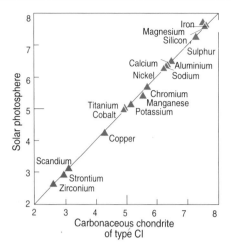

CHONDRITES

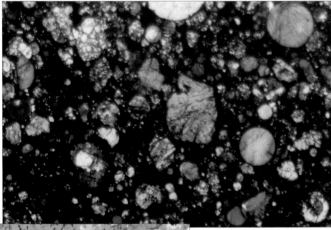

■ Left: comparison between the abundances of the major chemical elements in type CI chondrites and those measured in the solar photosphere (log scale). All the points fall on the diagonal line (slope 1, passing through the origin), which means that the abundances are identical in these two environments. The CI chondrites, the most primitive of all meteorites, have a composition representing the solar system in its entirety.

■ Right: transmitted-light view of a thin section of the Sainte Rose chondrite (x 10), showing that it is formed from an accumulation of chondrules.

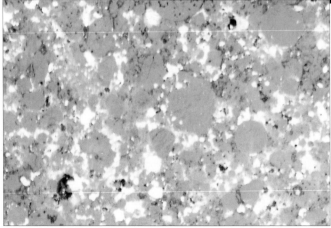

■ Left: reflected light view of the same area, showing the abundance of opaque minerals: metal and sulphides (white).

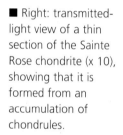

DIFFERENTIATED METEORITES

■ Right: comparison between the abundances of the major chemical elements in the Bouvante achondrite and those measured in the solar photosphere (log scale). Unlike the case of chondrites, most of the points are far from the diagonal line, which means these differentiated rocks have compositions very different from those of the sun (and the solar nebula). Eucrites, corresponding to the crust of a differentiated asteroid, are poor in elements which tend to enter the core (nickel, cobalt...) and rich in elements incompatible in mantle olivine (calcium, strontium...)

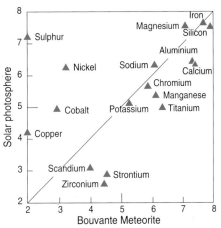

■ Above: thin section of the basaltic eucrite Juvinas, seen in cross-polarised light (× 10). Laths of feldspar (a calcium aluminium silicate) appear white, grey or black, depending on the orientation of the crystals sectioned; they are embedded in big coloured plates of pyroxene (an iron-magnesium-calcium silicate). This texture results from crystallisation of a silicate melt. There are no metallic grains.

■ Right: thin section of an olivine-rich terrestrial basalt (x 10). The same minerals are found again, but here there are large olivine crystals. Feldspar is less abundant and glass remains, indicating rapid crystallisation.

■ Above: thin section of a feldspathic lunar basalt (× 10). We see a similar texture to that of the picture on the left, although the feldspar laths are more abundant.

DIFFERENTIATED METEORITES

■ Right: polished slice of Peña Blanca spring, an enstatite achondrite (or aubrite). We see that the rock is made up of big grains of pyroxene (magnesium silicate) in a matrix of finer debris.

13 cm

■ Left: polished slice of the Springwater pallasite. The rounded amber-coloured olivine crystals are included in Fe-Ni.

■ Right: polished slice of the Guin iron containing, at this level, four dark nodules of iron sulphide, surrounded by a crown of iron phosphide. After polishing, a light chemical etch shows a host of criss-crossing bands, which correspond to the separation of phases poor and rich in nickel during the slow cooling of an initially homogeneous alloy. The width of the bands allows us to calculate the cooling rate of the alloy: it is of the order of 1 °C per million years. From this we deduce that the mass of iron from which the meteorite came was

13 cm

at the centre of an asteroid several hundred kilometres in diameter. Cooling rates this slow are impossible at the surface of the Earth, which is why the structures, so-called Widmanstätten patterns, are characteristic of meteorites.

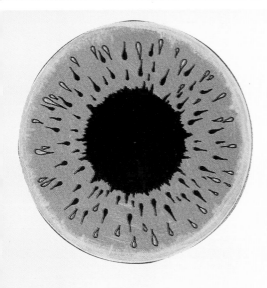

After differentiation

– **B**: if the sources of internal heat are sufficient to cause partial melting, differentiation happens: molten metal and sulphide migrate towards the centre to form the core (the origin of iron meteorites, we think). Silicates melting at low temperature produce a low density magma which migrates towards the surface where it erupts (and where we get those achondrites which are basalts).

■ Right: schematic cross section of the Earth showing the different superposed layers: core, mantle, crust and atmosphere (thicknesses not to scale).

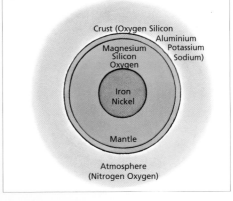

■ Left: the Pacaya volcano, in Guatemala. The internal heat of planets is much more quickly lost into space if they are small. Differentiated asteroids cooled and stopped evolving in the first few hundred million years of the solar system. The Moon became extinct three billion years ago. Some bigger planets, like the Earth, are still geologically active today: volcanism is linked to the persistence at depth of fluids and melted rock.

METAMORPHISM IN CHONDRITES

The sources of heat which affected chondritic asteroids were insufficient to induce partial melting. They nevertheless left their mark on the chondrites, by inducing a recrystallisation of the minerals, and a progressive obliteration of the textures, and notably of the chondrules. We call these "metamorphic" transformations. Unlike the terrestrial case, metamorphism in chondrites is strictly thermal and occurs in the absence of water. It is not linked to burial because internal pressures are low.

 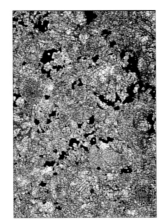 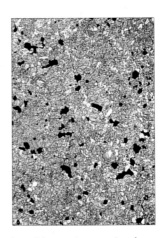

■ Above: transmitted light views of thin sections arranged in order of increasing grade of metamorphism (x 12). On the left (Mezö Madaras), the chondrules are very sharp. Some have mantles of opaque minerals (metal and sulphide), and the fine matrix which fills the interstices also appears black. In the centre (Monte das Fortes), the chondrule margins become hard to discern in places. There is no more fine matrix, and recrystallisation of opaque minerals is also visible: they have more compact shapes and tend to be grouped in the interstices between the chondrules. Finally on the right (Saint-Michel), chondrule boundaries are totally obliterated. Seen in the microscope, the rock has a compact texture. Metal and sulphide have equant forms, often with planar boundaries.

The metamorphic transformations in chondrites can hinder the reading of primordial processes predating parent asteroid formation. In this regard, the usually little metamorphosed carbonaceous chondrites constitute a precious resource. Unfortunately, their minerals have usually suffered a hydrothermal alteration, linked to the movement of fluids (especially water) towards the surface of their parent bodies. In fact, we do not know any which have come to us intact, as they were when formed by accretion in the solar nebula.

COLLISIONS AND BRECCIA FORMATION

The numerous collisions between asteroids or fragments which happened especially nearer the formation of the solar system have left their mark at every level in meteorites. Many rocks have a brecciated structure, which is to say they are the result of a reaccumulation of fragments from one or many rocks after impacts.

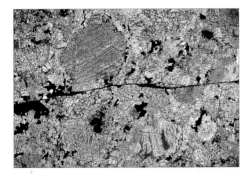

■ Above, the Aïr meteorite(x 9): thin section of a moderately shocked chondrite, seen in transmitted light. A shock vein crosses the field of view: it is a slip plane cut at right angles. The extent of displacement is of the order of a millimetre, as can be judged from the chondrule cut in two at the right of the image, with the two halves displaced from one another along the slip plane.

■ Below: thin section of the Jelica chondrite (× 9). We see that this rock formed by accumulation on a microscopic scale of often angular fragments (chondrule debris, notably), not particularly related to each other.

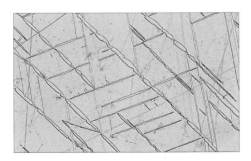

■ Above: Neumann bands in the Sancler Landia iron (x 120). These mechanical twins form during shocks of less than 130 kilobars. We see here lamellae oriented in three directions. The symmetry of the metal in fact allows twelve different orientations.

15,5 cm

■ Above: the Chinguetti meteorite. The meteorites of this type (mesosiderites) are the result of a mixture of iron and silicates. The generally accepted hypothesis is that they were formed during the impact of a giant iron projectile on a basaltic achondrite parent body.

THE PRINCIPAL METHODS OF STUDYING METEORITES: MICROSCOPY

The methods used for studying meteorites are the same as those applied to terrestrial rocks. Apart from traditional chemical analysis, the most classic methods use the optical microscope (transmitted and reflected light) and the electron microscope (scanning and transmission), as well as the electron microprobe. All these techniques allow us to obtain complementary information concerning the nature of the minerals observed, their crystal structure and their chemical composition.

■ Right: sketch of the principle of the scanning electron microscope and the electron microprobe. An electron beam bombards the sample. We can observe directly an electronic image which gives, depending on the detector, either topographic or mineral compositional information. We can also analyse x-rays emitted from the bombarded region. Their intensity at different wavelengths, compared to what is obtained on standards of known composition, allows us to deduce the composition of the sample: that is the principle of the microprobe.

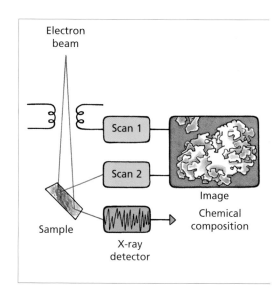

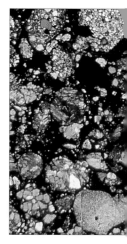

Meteorites: Rosetta stones for astrophysicists

In 1927, when the astronomer Henry Russell makes the first spectroscopic analyses of the Sun, he immediately thinks of comparing them with chemical abundances measured in meteorites. The idea that chondrites best represent the material composing the solar system is established little by little. In 1956, Hans Suess and Harold Urey show that the relative abundances of atomic nuclei making up this matter are not randomly distributed, but follow a few simple rules. To understand this distribution, it is necessary to consider that different nuclear processes give rise to different categories of nuclei. As early as 1957, key studies show that nucleosynthesis of the elements happens for the most part inside stars, and identify the broad lines of these different processes: nuclear astrophysics is born. Today still, and especially since 1987, discoveries made in meteorites are fundamental for astrophysicists: the identification and analysis of "presolar" grains allows us, indeed, to study in detail reactions which happened in different kinds of stars.

■ Below, pages 62 and 63, from left to right: three views (x 16) of the same field of view of a chondrite (St-Mary's County) in ordinary transmitted light (A), between crossed polars (B), and in back-scattered electrons (C). Ordinary light shows the outlines of the chondrules and the structure of the rock better. The colours of the minerals in cross-polarised light indicate their nature: the brightest colours (blue, purple, orange) indicate olivine. In the electron image the minerals (and therefore the chondrules) are more or less white according to whether they are richer or poorer in iron. This technique allows us to see whether olivine crystals are zoned, the normal pattern being rich in magnesium in the centre and rich in iron at the rim.

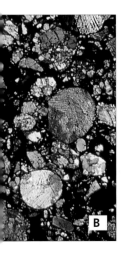

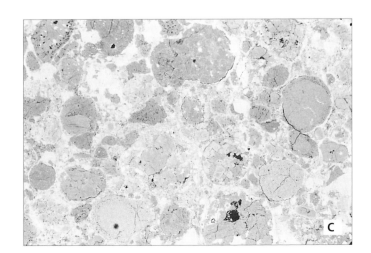

THE PRINCIPAL METHODS OF STUDYING METEORITES: ISOTOPIC ANALYSIS

The mass of atoms is concentrated in their nucleus, made up of protons (charged positively) and neutrons. A "cloud" of electrons (charged negatively) maintains neutrality and is responsible for the chemical properties. When the number of protons in the nucleus changes, the number of electrons changes too and we obtain a new chemical element. However, if the number of neutrons in the nucleus varies only the mass of the atom is changed (and not its chemical properties) and we do not change elements. We use the name isotopes for nuclei of different masses making up the same element. These elements are normally found in virtually constant proportions. For example, chlorine (17 protons) has an atomic mass of 35.5 units, as it is made up of ¾ of the isotope 35 (18 neutrons) and ¼ of the isotope 37 (20 neutrons). Small variations around this value can be produced as the result of physico-chemical processes in which mass plays a role, such as evaporation which tends to concentrate the heavy isotope in the residue. We speak then of mass fractionation, and we verify, when there are more than two isotopes, that the relative variations in their abundances are proportional to their differences in mass. Such measurements are done with the aide of a mass spectrometer, in which the

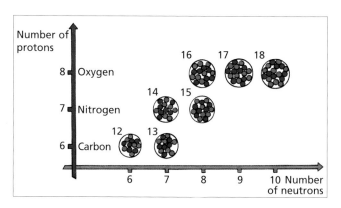

■ Above: on this schematic representation of atomic nuclei making up several elements, the protons are shown in blue and the neutrons in red. The mass of a nucleus (atomic mass) is equal to the sum of its number of protons and neutrons. Most chemical elements are made up of several isotopes having the same number of protons but a different number of neutrons. Thus oxygen (8 protons) includes three isotopes of mass 16, 17 and 18, according to whether they have 8, 9 or 10 neutrons. As a general rule, the different isotopes are present in constant proportions. When these proportions are different, we speak of isotope anomalies.

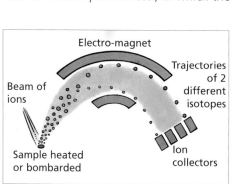

■ Left: sketch of the principle of the mass spectrometer.

atoms are emitted by the source in ionised form (charged by the loss of an electron), then accelerated by an electric field and deviated by a magnetic field. The radius of curvature of the trajectory followed by the ions depends on their mass, which allows us to separate the different ions by modifying the magnetic field or the location of the ion detectors. These measurements are usually expressed in the form of isotope ratios, relative to a reference isotope. To make analyses in a mass spectrometer, it is generally necessary to dissolve the specimen beforehand and proceed to a chemical separation of the elements to be analysed. In the case of the "ion probe", however, the whole sample, is bombarded by a beam of primary ions, and emits directly the ions which are analysed.

In certain cases, variations in isotope ratios can occur, which are not linked only to mass fractionation. Their significance is explained on pages 102 to 121. The major variations in the isotope ratios of oxygen which exist from one meteorite class to another reinforce the mineralogical and chemical classification: they indicate that we are dealing with objects coming from distinct "reservoirs". While the existence of reservoirs differing in oxygen gives us an almost infallible way to classify meteorites and to demonstrate certain links possibly existing between different classes, the fundamental cause of these reservoirs is still a subject for debate. It is possible

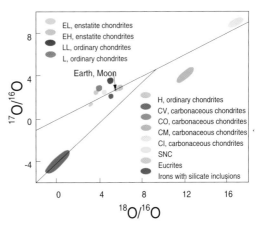

■ Above: distribution of the three isotopes of oxygen in meteorites. The values represent relative differences (expressed in thousandths) from the reference value (oxygen in sea water).

that they are due to a late heterogeneous seeding of the solar nebula by material coming from a star older than the sun and containing pure oxygen 16. But we can also consider the effect of chemical reactions such as are produced in molecular clouds.

Key words: **achondrite • asteroid • basalt • breccia • carbonaceous chondrite • chemical separation • chondrite • chondrule • crust of the Earth • crystallisation • differentiation • electron microprobe •** (scanning, transmission) **electron microscope • eucrite • fusion crust • hydrothermal alteration • ion • ion probe • isotope anomaly • isotope ratio • light** (natural, cross-polarised, reflected) **• mass fractionation • mass spectrometer • matrix • metamorphism • meteorite • Neumann bands • nodule • nuclear astrophysics • nucleosynthesis • nucleus • olivine • oxidised • planet • pyroxene • reduced • refractory inclusion • shock vein • shocked • silicate • solar photosphere • thin section • twin • white inclusion • Widmanstätten pattern •**

CLASSIFICATION OF METEORITES
CHONDRITES

CARBONACEOUS CHONDRITES	**CI**	*example : Orgueil*
	CM	*example : Murchison*
	CO	*example : Ornans*
	CV	*examples : Allende, Axtell*
	CK	*example : Karounda*
	CR	*example : Renazzo*
	CH	*example : ALH85085*

R	*example : Rumuruti*

ORDINARY CHONDRITES	**LL**	*examples : Saint-Mesmin, Douar Mghila*
	L	*examples : L'Aigle, Mezö Madaras*
	H	*example : Flandreau*

AL	*examples : Acapulco, Lodran*

Brachinites	*example : Brachin*

ENSTATITE CHONDRITES	**EL**	*example : Eagle*
	EH	*example : Saint-Sauveur*

CLASSIFICATION OF METEORITES
DIFFERENTIATED METEORITES

Irons *examples : Staunton, Tamentit*		*IRON METEORITES*
Irons with silicate inclusions		
Pallasites *example : Springwater*		

Mesosiderites *example : Chinguetti*	
Eucrites *example : Bouvante*	*ACHONDRITES*
Diogenites *example : Tatahouine*	
Howardites *example : Le Teilleul*	
Angrites *example : Angra dos Reis*	
Ureilites *example : Novo Urei*	
SNC (Mars ?) *examples : Chassigny, Zagami*	
Basalts and lunar brecccias	
Aubrites *example : Peña Blanca Springs*	

Little planets

Where do meteorites come from? Their study in the laboratory, coupled with astronomical observations, shows that the majority are derived from a population of small planets lying mainly between the orbits of Mars and Jupiter.

Since ancient times, astronomers knew of seven celestial bodies which move relative to the stars: the Sun, the Moon, and five planets (from a Greek word meaning wandering star) visible to the naked eye: Mercury, Venus, Mars, Jupiter and Saturn. We can add the comets, whose movements were however not clarified until about 1700 by Edmond Halley.

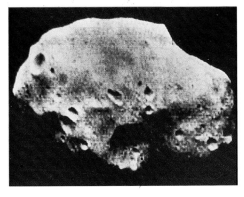

Above: Phobos (28 km by 20 km) is one of the two satellites of Mars (the other being Deimos). Its low density (1.9 g/cm³) suggests that it corresponds to a (porous) carbonaceous asteroid formed at a greater distance from the Sun and later captured by Mars.

Left page: the asteroid belt imagined by a contemporary painter, with an asteroid in the foreground. Even though it is populated with tens of thousands of little bodies, the effective rarity of these, in the vast space between Mars and Jupiter, is well represented in this image.

SEARCHING FOR A MISSING PLANET

In the second half of the 18th century, the discovery of an approximate empirical relationship between the dimensions of planetary orbits, the "law" of Titius-Bode (the names of two astronomers who popularised it), suggests that a planet is missing between Mars and Jupiter, at a distance of about 2.8 times the average Earth-Sun distance (i.e. 2.8 astronomical units or 2.8 AU for short) from the Sun. In 1781, William Herschel the British astronomer of German origin discovers a new planet beyond Saturn. This receives the name Uranus. Its orbit is again consistent with the Titius-Bode law. The observers gear up to identify the missing planet. Instead of a single object of respectable dimensions, more than 300 little planets are discovered by astronomers in the course of the 19th century. The biggest, Ceres, has a diameter less than 1,000 km. With the coming of systematic searches thanks to photography, after 1891, the number of objects discovered grows rapidly: we now number them at more than 30,000, of which a little more than 7,000 follow orbits which have been calculated with precision. The majority are concentrated at distances between 2.2 and 3.3 AU from the Sun, forming what we call the asteroid belt.

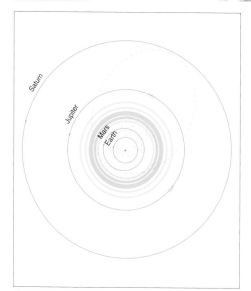

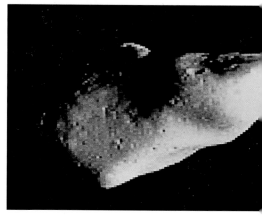

Above: between the orbits of Mars and Jupiter tens of thousands of asteroids show stable orbits around the Sun, making up the asteroid belt. Other asteroids (less numerous) have highly elliptical orbits which cross those of the inner planets or approach that of Jupiter, on which we note the two groups of Trojan asteroids.

THE CONTRIBUTIONS OF SATELLITES AND RADAR

Space observatories now make a major contribution to the discovery and study of these small bodies. In 1983, the IRAS satellite (InfraRed Astronomical Satellite) allowed us to establish a map of the asteroid belt in the infrared: its successor ISO (Infrared Space Observatory), continued this task from 1995 to 1998, with a better sensitivity. A number of asteroids were studied at the beginning of the nineties by the European astronomical satellite Hipparcos, and others by the Hubble Space Telescope. Gaspra and Ida were flown-by and photographed, in 1991 and 1993 respectively, by the American probe Galileo, during its journey towards Jupiter. NEAR is spending 2000 in orbit around Eros.

Finally, asteroids which pass close to Earth can be studied with the aid of big

Above: Gaspra, a silicate-dominated asteroid of very irregular form (19 km by 12 km), probably results from a recent collision (about 200 million years ago),based on the age of its weakly cratered surface.

At top: the asteroid Ida (58 km by 23 km) is the product of the fragmentation of a big asteroid, whose disruption dates back 1 to 2 billion years, based on its highly cratered surface. It has a spectrum of the silicate-dominated type, but fresh ejecta resemble ordinary chondrite. Its satellite, Dactyl (right side of image) measures 1.5 km.

radio telescopes using radar, and we have been able to observe the form of several of these bodies.

RESONANCES AND ASTEROID FAMILIES

The movement of an asteroid, like that of a bigger planet, is governed by Kepler's laws: its orbit is an ellipse with the Sun occupying one of the foci; its orbital velocity depends on

Are asteroids the debris of a planet which exploded?

While searching for a planet which seemed to be missing between Mars and Jupiter, the Italian G. Piazzi discovered the asteroid Ceres, at 2.77 AU from the Sun, on January 1st 1801. But other asteroids would soon be observed, and we now know more than 30,000. Are these little bodies fragments of the "missing" planet, which might have exploded, or are they really residual planetesimals which were unable to accrete to form this planet? The study of meteorites allows us to show that the second hypothesis is correct. The most important argument is that of the oxygen isotope signatures: if meteorites came from the same planet, they would all have identical oxygen signatures, and this is obviously not the case.

its distance from the Sun; and finally there is a relationship between the semi-major axis of its orbit and the time it takes to complete one revolution around the Sun.

In 1857, the American astronomer Daniel Kirkwood studies the distribution of semi-major axes of asteroid orbits as a function of the distance from the Sun. He then discovers that this distribution is not uniform. Certain zones are practically empty: we now call them the Kirkwood Gaps. They correspond to orbits where the period of revolution around the Sun makes a simple whole-number ratio with that of Jupiter. From a physical point of view, this corresponds to (gravitational) resonances with Jupiter: for example, the 3/2 resonance is that for which an asteroid makes three revolutions around the Sun while Jupiter makes two. The more the two numbers are small, the more the resonance is "strong". We will see that resonances play an essential role in the transfer of asteroids (or fragments of asteroids) into orbits which bring them into the vicinity of the planets close to the Sun, in particular the Earth.

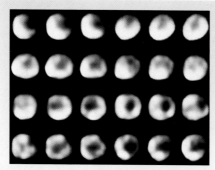

These 24 snapshots of Vesta were taken by the Hubble telescope. They show that the third asteroid by size (501 km) does not have a totally spherical form. We think that it is perhaps the parent body of certain differentiated meteorites.

Asteroids and meteorites

To grasp how they are connected, we can compare the spectra of light reflected by various types of asteroids studied by telescope with those of meteorites crushed in the laboratory. Indeed we think that the surfaces of asteroids, free of atmosphere and tilled by ceaseless impacts, must resemble such a blend of dust and mineral fragments. As early as 1970, the striking resemblance between the spectra of the asteroid Vesta and the eucrites confirmed the validity of this approach. So we have copied the classification of asteroids (carbonaceous or differentiated) from that of meteorites. Since the eighties, however, one question has excited astronomers and meteoriticists: which asteroids are the source of the three classes of ordinary chondrites which represent 75% of the falls? For some, no asteroid bigger than 40 km matches the spectra of the latter; others maintain that, because a third of exotic fragments trapped in 4.56 billion years in meteoritic breccias are fragments of ordinary chondrites, the three parent asteroids are not rare objects and must be there in the main belt. A recent response from the astronomers recognises that a third of the asteroids with silicate-dominated spectra "could" be the source of ordinary chondrites. Now NEAR x-ray data suggest that Eros is an ordinary chondrite parent body.

In 1918, the Japanese astronomer Kiyotsugu Hirayama notices that three groups of asteroids show analogous orbital characteristics. He then assumes there exists a link between the members of each of the groups, to which he gives the name "families", and he suggests that these little bodies could be debris resulting from the fragmentation of a larger object. Today, we recognise about twenty asteroid families, and there is no longer any doubt that they result from fragmentation following a collision between two small planets.

The distribution of average distances from the Sun for about 4,000 asteroids in the main belt is irregular. The majority of asteroids are spread between 2.1 and 3.3 AU. The "gaps" in the distribution are due to "resonance effects" with the period of Jupiter (indicated by red arrows).

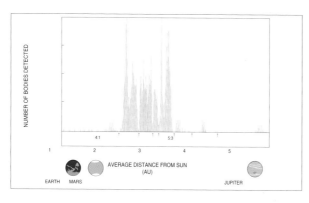

NUMBER OF BODIES DETECTED

4:1   5:3

AVERAGE DISTANCE FROM SUN (AU)

EARTH   MARS   JUPITER

ASTEROIDS WHICH GRAZE THE EARTH

As long as the members of a family stay far from a resonance, they travel soberly around the Sun in the asteroid belt. But, once some of them approach one of these zones too closely, they are either ejected towards the outskirts of the solar system, or sent towards the Sun. In the latter case, if they do not crash into the Sun, they can be trapped by the inner planets (including the Earth). Then we give them the name NEA (Near-Earth Asteroids, or asteroids which graze the Earth). Some NEA do indeed encounter the Earth, fall to its surface and sometimes excavate spectacular impact craters, like Manicouagan (100 km in diameter) in Canada, or of smaller size like the famous Meteor Crater in Arizona. NEA are responsible for the majority of meteorites.

How can such objects reach us? Let's take the case of the 3/1 resonance. The corresponding orbit will be progressively modified by the attraction of Jupiter. After a

At top: asteroid 4139 Toutatis (NEA class), discovered in 1989, passes in the night sky.

Above: radar image taken on December 8th 1992 when Toutatis (whose orbit crosses that of the Earth) was 2.5 million kilometres from our planet. Toutatis seems to be formed of two fragments (2.5 km and 4 km) held together by gravity

few million years, it will be transformed into an ellipse so elongated that it will cross the orbit of Mars and even that of the Earth or Venus. The asteroid Toutatis, discovered in 1989 at the Côte d'Azur observatory, is today in this situation and passes close to Earth about every four years. Numerous other resonances are just as efficient at modifying asteroid orbits. We should also note that repeated passages close to the inner planets have the effect of making the NEA orbits very chaotic. It then becomes very difficult, almost impossible, to predict their long term future. Thus we know the orbit of Toutatis quite well for the next thousand years, but beyond that the calculations are very uncertain.

Among the arguments which suggest that the NEA are really fragments of main belt asteroids, we can cite three: first of all, their albedos (proportion of light reflected by their surfaces) are analogous to those of asteroids in the main belt; next, the surface mineralogical compositions, deduced by absorption spectrometry, are similar; finally their dimensions never exceed about 40 kilometres, which is much smaller than those of the big asteroids in the main belt,

indicating that we are dealing with fragments.

METEORITE PARENT BODIES

There is a link between the history of meteorites and that of asteroids. Everything began in the medium of the nebula, from which the planets which surround the Sun are derived, by accretion. Between Mars and Jupiter, the resonances due to the great mass of Jupiter have prevented the formation of a significant planet, and have since unleashed collisions between asteroids by clearing the Kirkwood Gaps.

Meteorites which come from Mars?

The basaltic differentiated meteorites (eucrites) formed very early in solar system history, between 4.56 and 4.45 billion years ago. But a little group of 19 achondrites, the SNC (shergottite, nakhlite, chassignite) and related meteorites, stand out by having ages mainly less than 1.3 billion years. Their "youth" cannot be attributed to late differentiation in asteroids; in fact asteroids with their small sizes, cooled very quickly. The SNC meteorites all carry the same oxygen isotope signature, indicating that they come from the same parent body. They are more oxidised than eucrites, and also show isotope ratios for certain rare gases (as well as nitrogen and hydrogen) similar to those measured in the Martian atmosphere. Their Martian origin thus appears probable. The Moon has also sent us about 21 meteorites.

ALH 81005 is the first meteorite of lunar origin discovered in the Antarctic. It is a regolith breccia, containing feldspar-rich clasts (white), and a matrix, made up of mineral fragments and abundant brown glass, very enriched in gas implanted by the solar wind.

The Shergotty meteorite (in ordinary light, on the left) is a basaltic rock of probable Martian origin, made up of feldspar (clear) and pyroxene (brown), which suffered a strong shock which made the feldspar amorphous. In cross-polarised light (on the right), the black areas are feldspar, turned to glass by the shock.

These orbits of Earth-crossing NEA represent less than 5% of the asteroids bigger than 1 km which cross the Earth's orbit.

Two specimens of the meteoritic breccia Djermaia, from the parent asteroid of the H chondrites. At the top the light fragments have been rounded by the surface erosion (micrometeorite flux) which operated for nearly 30 million years, while the black matrix made up of fine grains was enriched in gas from the solar wind. The piece below comes from a depth of 30 to 50 cm where the erosion processes are much weaker, where the light matrix is little enriched in solar wind and where the more abundant lithic fragments have almost all retained their angular shape.

Could an asteroid run into the Earth?

More than a hundred asteroids of kilometre size which cross the Earth's orbit are catalogued, and this represents perhaps 5% of the total of these fragments. There is thus a certain risk of a collision of an object a few kilometres in size with our planet. This could provoke a climatic upheaval and extinctions of species, as in the past. Even a hundred metre fragment could do great damage, and objects of this size are certainly much more numerous still than those of kilometre size. In 1992, Toutatis passed the Earth at a distance of 3.6 million kilometres, and it seems that it will approach closer in a few decades. A radar study has shown that it is formed by the juxtaposition of two blocks, one 4 km, one 2.5 km: a collision with this object would have dramatic consequences. This is just an example, but the risk of an asteroid collision with Earth can now no longer be dodged, and scientists study the means to confront it.

In the part of the asteroid belt closest to the Sun, the asteroids have experienced the highest temperatures, resulting in metamorphism (at temperatures of 400 to 900 °C) or differentiation (temperatures over 1200 °C). At greater distances from the Sun, the asteroids have better preserved their primitive nature: their richness in volatile elements and especially water has moderated their heating (50 to 400 °C). In every case, the sources of heat may have been the young Sun and certain short-lived radioactive elements, like aluminium 26 for example. Thanks to collisions, some fragments have later fed the flux of meteorites. Fragments of undifferentiated asteroids have yielded chondrites. Fragments of differentiated asteroids have yielded different classes of meteorite, depending on their original position in the body they come from: fragments of the core yield irons, of the mantle pallasites and diogenites, and of the crust eucrites.

Key words: absorption **spectrometry** • accretion • achondrite • albedo • **asteroid belt** • **astronomical unit** • **breccia** • **comet** • **chondrite** • **crater** • **differentiation** • **diogenite** • **eucrite** • **focus** • **hydrosphere** • **impact** • inner **planet** • **iron** • **isotopic** signature • **Kepler's laws** • **Kirkwood Gaps** • **meteorite** • **NEA** • **orbit** • **pallasite** • **parent body** • **planetesimal** • **resonance** • **satellite** • **semi-major axis** • **spectrum**

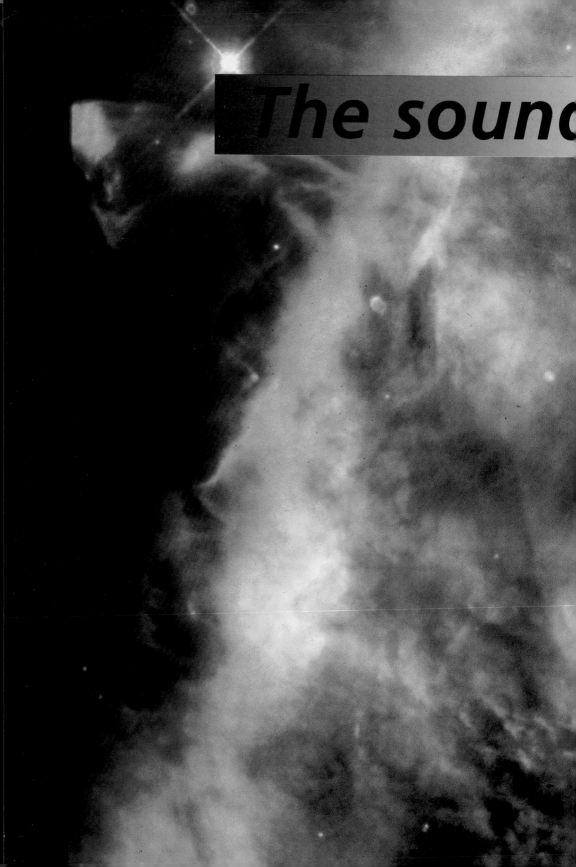

The sound

and the fury

Three decades of space exploration, astronomical observations and laboratory analyses of extraterrestrial material have revolutionised our vision of the Universe at every scale, and especially that of the world of planets. While our understanding of stellar evolution allows us to classify stars in families of similar objects, the planets in the solar system strike us by their astonishing diversity. However, they were all formed at the same moment – a little more than 4.56 billion years ago - in the same part of the Galaxy and from the same material. Where does their diversity come from?

THE SEARCH FOR OUR ORIGINS

The diversity of planets comes partly from the initial conditions for the evolution of the solar system: the composition of the original material, the primordial kinetics and dynamics. In other words, it is during the processes dating from the very formation of the solar system that the present characteristics of the objects which surround us were established. How can we identify the processes?

To decipher the conditions of origin of the solar system, we use two types of information. On the one hand, we can observe regions of star formation, elsewhere in the Galaxy, which we think are

Left page: in the vast Orion cloud of our galaxy, hundreds of stars in the process of forming, like our Sun at its birth, teach us directly about the first stages of evolution of our solar system, 4.56 billion years ago.

Panoramic view of our galaxy, the Milky Way: the long bright trail includes very bright stars and vast dark regions of clouds containing molecular gas and dust.

Left: view in ultraviolet of a prominence on the Sun.

Below: when they die, massive stars explode as supernovae, ejecting filaments of gas containing freshly synthesised heavy atoms into the interstellar medium.

comparable to that of the solar system at its birth: the Orion nebula is an example. On the other hand, we have in the solar system objects which have, in part at least, preserved intact the chemical compositions acquired during their formation. Such fossils can be found in the smallest bodies, cometary nuclei and asteroids: we can either try to explore them by space missions, or to analyse them in the laboratory, as meteorites and micrometeorites are derived from them.

FROM INTERSTELLAR CLOUDS TO STARS

Astronomical observations, and notably radio astronomy, have taught us that stars form in interstellar "clouds". These gigantic structures of gas and dust, which represent hundreds of thousands of solar masses, are visible to the naked eye in our Galaxy. They form dark zones in the Milky Way, so dense that they prevent us from seeing the stars behind them. The interior of these clouds, protected from stellar ultraviolet radiation, is the site of a very efficient chemistry for the synthesis of molecules. Hydrogen is present mainly in the form of the molecule

H_2, and carbon, though mostly combined as CO, is also found in organic chains, some of which can be very large. In fact, when clouds collapse on themselves under the effect of gravity, collisions between atoms and molecules become more and more frequent, giving rise to more and more complex molecular structures. The material forming stars and planetary bodies must then contain, besides refractory interstellar grains, a very great variety of organic chemicals.

The instability of these molecular clouds leads to their fragmentation. Each fragment can later collapse under its own weight to give birth to a star. This collapse of molecular clouds happens on a scale of a few million years. That represents only a thousandth of the age of galaxies, a hundredth of their period of revolution. Once formed, stars evolve and die after a time depending on their mass: the more massive the star, the faster it evolves. A star like the Sun, formed 4.566 billion years ago, has only burned about half of its fuel. A star twenty times more massive will explode in a supernova after only a few million years. Now this time

The solar system

The solar system consists of a star, the Sun, and all the bodies located in the region of space where the attraction of the sun is greater than that of other stars: nine main planets and their satellites, a multitude of small planets (or asteroids) and comets, meteorites and interplanetary dust.

The nine main planets are split into two groups: those closest to the Sun (Mercury, Venus, the Earth and Mars, to which we can add the Moon) are relatively small but dense, have evolved profoundly since there formation and have differentiated structures. Those which are further from the Sun (Jupiter, Saturn, Uranus and Neptune) are distinctly bigger but less dense. These are almost entirely gaseous planets whose atmosphere, mainly hydrogen and helium, has evolved very little since their formation. Pluto, the planet furthest from the Sun (with a very elliptical orbit, on average forty times further from the Sun than the Earth), is an icy body smaller than the Moon.

The majority of the asteroids are concentrated between the orbits of Mars and Jupiter. We think that the comets form a vast halo (the Oort cloud) at distances up to several tens of astronomical units from the Sun.

is comparable to that of the contraction of the original cloud: thus massive stars form, evolve and die in the heart of a cloud. We even think that, because of the shock wave that it emits, the explosion of a supernova can cause the collapse of cloud fragments leading to the appearance of other protostellar nebulae, and seed them with newly synthesised elements at the same time. Certain of these elements are radioactive and their decay within planetary bodies later formed around the new star is one of the sources of heat causing metamorphism and differentiation of these bodies. The Sun itself at its birth would have been in the vicinity of one or several supernovae, as witnessed by the "isotopic anomalies" discovered in certain meteorites.

STARS AND PLANETS

The formation of stars with masses comparable to that of the Sun is however not necessarily linked to the explosion of massive stars. There exist numerous sites, for example in the constellation of Orion, where we have recently detected a very large population of small young stars without observing massive stars: it seems that the collapse of gigantic molecular clouds yields zones containing a host of stars like the Sun. High resolution images of these stars show that more than a third are surrounded by rotating circumstellar haloes which, in a few

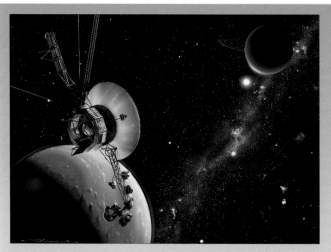

Exploration of the solar system

In thirty years of space exploration, all the planets of the solar system except Pluto have been flown over by at least one space probe, sometimes more than ten, as in the case of the Moon, Mars and Venus. These American and Soviet probes, equipped with instruments developed with the

Artist's view showing the Voyager 2 probe in the vicinity of Triton whose surface appears to be composed of methane and nitrogen ices; behind, Neptune and its fine rings.

participation of several tens of nations, have revealed a diversity of an unsuspected richness. Besides the giant planets, their rings and satellites, which Voyager encountered, some small solar system bodies have also been observed: satellites (Phobos and Deimos) were surveyed by the Viking and Phobos missions to Mars, and asteroids (Gaspra, Ida) were encountered by Galileo, en route to Jupiter. The American probe NEAR Shoemaker recently flew by the asteroid Mathilde and in 2000 went into orbit around the asteroid Eros for an extended study.

The nucleus of comet Halley was visited by five space probes: two Japanese, two Soviet (Vega 1 and 2) and one European (Giotto). NASA's Stardust mission is on the way to comet Wild 2 to collect samples of comet dust, which it will return to Earth in 2006. The Contour mission will fly by three comets beginning with Encke in November 2003. The European Space Agency will launch Rosetta to rendezvous with the comet Wirtanen, in 2012. The probe will send two laboratories to land on the nucleus, to analyse its composition, but will itself stay in a low altitude orbit, to follow the evolution of cometary activity for several months, until its closest approach to the Sun. Currently, international co-operation (United States, Europe, Russia, Japan) is important for planning missions to the Moon and Mars.

hundred thousand years, will flatten: in these disks of rotating matter, protoplanetary objects could be born. The latter hypothesis is supported by numerous observations of small periodic variations in the movement of nearby stars, considered as sensitive indicators of the presence of planets around these stars. No doubt continued observations by the Hubble space telescope and infrared satellites like ISO will add considerably to our understanding of this stage of the formation of stars and planetary systems.

Mars panorama, photographed by the Viking 1 probe, shows a reddish desert which winter covers with frost (carbon dioxide and water ice).

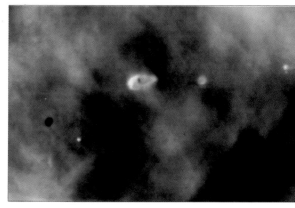

Around some stars forming in the Orion complex, we see a disk of matter; it could be a protoplanetary cloud, in which case we are observing the formation of a stellar system.

The images from Hubble seem to show that the central star ignites even when the matter surrounding it is not yet accreted into protoplanetary objects of great size. This is consistent with the idea that nebulae reach high temperatures in their central regions, to the point where most grains are evaporated and incorporated into the gas: protoplanetary material would recondense later from a very well homogenised mixture. In the case of the solar system, local inhomogeneities have been preserved however.

One of the major results of the last thirty years is that we have shown, by the discovery of "isotopic anomalies" in certain meteoritic minerals, that the solar nebula was not perfectly homogenised: its presolar memory has not been totally erased. Quite the opposite: there exist grains, synthesised in the atmosphere of presolar stars, which have survived the formation of the Sun and the phase of accretion of its planetary companions. The identification of such grains allows us to track the history and geography of the nascent solar system.

PLANETS AND SMALL BODIES

To understand what the study of meteorites yields, just think about the origin and evolution of their parent bodies. Why have certain asteroids preserved their initial physico-chemical properties, while we know that a planet like the Earth has lost all trace of its distant past?

The rocks of Earth were not formed at its origin, in the state in which we now find them: multiple transformations have modified their compositions. If the Earth is still active from a geological point of view, it is because, like all the other planetary bodies, it contains a source of energy principally in the form of the long-lived radioactive elements: uranium, thorium and potassium. The power liberated is a function of the total quantity of these elements, and so of the volume of these objects. The bigger they are, the more heat they produce. The smaller they are, the more radiative loss will allow them to stay cold. So the Earth achieved a central temperature higher than that of Mars, which in turn reached a higher one than the

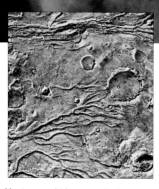

In the Martian past, higher atmospheric pressure must have allowed water to exist in a liquid state, as demonstrated by the fluvial network – dry today – visible in the image.

The Magellan mission has allowed us to determine the topography of Venus, and to construct Venusian panoramas by computer. We see tectonic structures, volcanoes and impact craters. We are not certain however that the planet is still geologically active.

The origin of the Moon

While the Apollo missions radically changed our understanding of the Moon, and the first billion years of inner solar system evolution, there was one question that the analysis of lunar samples did not entirely resolve: that of the origin of the Moon. Previously, three different models had been proposed: either the Moon was captured from a nearby orbit by the influence of the Earth's gravity; the Earth and the Moon accreted simultaneously in the same dust disk and stayed in orbit one around the other; or the Moon originated by fission of the mantle of the Earth, supposedly initially liquid and rotating rapidly.

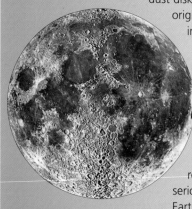

Each of these hypotheses faces serious difficulties. The first two do not explain the great difference in density and composition between the Earth and the Moon, the latter containing very little of the elements present in the Earth's core (iron and nickel); the second hypothesis, furthermore, does not take into account the present angular momentum of the Earth-Moon system. The third resolves the problem of iron content, but also faces the serious problem of the initial angular momentum of the Earth.

The surface of the Moon is peppered with craters mostly dating from the first billion years of its history. The lunar maria, which result from an outpouring of lava, demonstrate extensive internal activity between three and four billion years ago.

Computer simulations, coupled with the nature of lunar samples, favour a fourth model, a hybrid of the three earlier ones: the Earth suffered a giant impact with an object comparable to Mars, about 4.5 billion years ago. Part of the material ejected would have been from the Earth's mantle, in liquid and gaseous form, and would have been placed in Earth orbit. The considerable heating due to the impact would have vaporised the most volatile elements, which escaped reaccretion into the Moon we know today.

Miranda (diameter 480 km), the innermost satellite of Uranus, made of ices which melt at very low temperatures, shows very diverse terrains (cratered terrain, faults, and mountains-valleys), the latter possibly due to the tidal action of Uranus.

View of the scars caused on Jupiter in July 1994 by the fall of fragments of the comet Shoemaker-Levy 9. The impact of comet fragments produced much bigger effects than predicted. For example, the scar at lower left is about the size of the Earth.

Moon. Once these temperatures were reached, they did not last: by the very nature of the energy source which comes from radioactive decay, the efficiency of this source diminishes with time. Finally, the biggest objects attain greater degrees of activity, which last longer. With its earthquakes and volcanism, the Earth shows that it is still active: Mars, inactive for several hundred million years, had previously developed impressive forms of tectonic activity. The Moon went extinct about two billion years earlier, only about two billion years after its formation. The filling of the lunar basaltic maria is the largest scale phenomenon linked to its internal activity.

Apart from the energy produced by radioactive decay, the initial heating of planetary bodies was due to the liberation of gravitational energy (if their formation was rapid enough), as well as the energy released during the numerous collisions which marked the last stages of the accretion process.

Thus it is among the small bodies of the solar system that we can hope to find objects having suffered the least changes linked to global heating. In the asteroid belt, it is the objects furthest from the Sun, rich in volatile elements and very hydrated, which have experienced the weakest metamorphic alteration: they are the source of carbonaceous chondrites. Similarly cometary nuclei have escaped every kind of heating. These two classes of small, undifferentiated bodies constitute the source of the most primitive samples of the solar system.

We owe the survival of the asteroids, a reservoir of rocky fragments and blocks of varied dimensions, anything up to hundreds of kilometres, to the accretion of a considerable mass of gas, leading to the formation of Jupiter. They move rapidly on crossing orbits, and their collisions supply debris, some in the form of meteorites, which remains one of the richest sources for evidence of the original accretion process.

Well beyond the orbit of Jupiter, where the temperature remained sufficiently low that "ices" of all kinds remained stable, most of the bodies are made up of these frozen molecules, the other grains present in the nebula. To a large extent, these ices contain complex organic molecules. Tens of billions of cometary objects thus grew in the solar nebula to kilometre sizes. Only a small number reached much greater sizes. It seems that, just as Jupiter had an effect on asteroids, the appearance of Saturn, Uranus and Neptune had the effect, because of the resulting gravitational perturbations, of expelling most of these comets to very great distances from the Sun (several tens of thousands of AU): today they populate a vast reservoir, "the Oort cloud", named for the astronomer who first suggested this idea. At such distances, the comets are only weakly bound to the Sun and the attraction of a nearby star is sufficient to modify the orbits of some of them. When they fall towards the Sun they become observable, thanks to their tails of gas and dust caused by the sublimation of the ices heated by solar radiation. Micrometeorites, which are in part derived from comets, are naturally considered very precious primitive specimens!

A LONG EVOLUTION

Within galaxies, which are permanent structures, matter evolves, stars appear, transform and disappear. The Sun is no exception to this rule: it is not a first generation star. Like the rest of the solar system, it has incorporated gas and grains derived from earlier stars. The solar nebula

Heat sources of small planetary bodies

Meteorites record early and varied thermal evolution of their parent bodies, principally asteroids: while some did not reach 100 °C, others exceeded 1,200 °C and experienced melting. The heat sources which transformed these planetoids have not been identified with certainty. They are not the same as those of planets like the Earth. The accretion of small bodies like asteroids results from low velocity encounters which release little heat. Long-lived isotopes (uranium, thorium and potassium) release energy too slowly: it is rapidly lost by radiation from the surface and therefore cannot heat the small bodies. Two possible processes have been considered:

Thin section of the eucrite Juvinas in cross polarised light. This rock was produced by crystallisation of a melt heated above 1,200 °C. Crystals of pyroxene are bright colours and feldspar white and grey. This basaltic lava was erupted on the surface of a differentiated asteroid which could have been Vesta.

– external heating of the small bodies, linked to an intense early activity in the Sun. The electric current induced by the flux of ions emitted by the Sun could have heated bodies with a size of a few tens of kilometres (by the Joule effect).
– internal heating by short-lived radioactivity (aluminium-26 for example), considered by H. Urey as early as 1955. In 1976, it was shown that this radioactive isotope had a sufficient abundance in certain chondrites to lead to rapid melting (in less than a million years) of ten-kilometre chondritic objects

Collecting micrometeorites

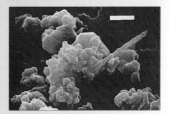

Cometary nuclei heated near the Sun vaporise at the surface, liberating grains trapped in the ice. The smallest grains are pushed by radiation pressure and become interplanetary dust. Those bigger than a few microns are concentrated in meteor swarms along the orbit of their parent comet. When the earth crosses these swarms, the biggest grains are heated in the atmosphere forming shooting stars; others fall to the ground or in the ocean where they can be collected in marine sediments and polar ice. They have also been collected high in the atmosphere using U2 planes, but quite recently we had the idea of collecting them in space, that is to say before they are transformed by heating. Even better: we can achieve specific collections, comet by comet, of this unique material, by putting collectors outside a space station, and exposing them in space only during the Earth's passage through a particular swarm. An experimental set-up of this type, developed in France at the Institut d'Astrophysique Spatiale, was installed in 1995 on the Russian station Mir. The extraterrestrial grains arriving at high velocity hit a material capable of slowing them down without destroying them, which is placed in sealed boxes, to limit contamination between space and the laboratory. After opening and exposure in space, the boxes are closed and recovered by astronauts, then returned to Earth, where the grains are individually analysed.

Scanning electron microscope view of interplanetary dust particle, showing sub-micron grains.

evolved simultaneously by chemical changes in a large fraction of the gas and the appearance of large planets. A complex series of collisions followed by accretion or fragmentation profoundly modified the spatial distribution and size spectrum of these objects. A certain population of "small bodies" escaped growth into planetary size objects, destruction by impact on a planet, accretion into the Sun, and finally expulsion from the solar system. These objects preserve today the memories of the multiple processes which have led to the solar system as we see it today.

key words: **accretion • angular momentum • asteroid • circumstellar halo • comet • contamination • differentiation • galaxy • impact • interstellar** (cloud, grain, medium) **• isotopic anomaly • metamorphism • meteor stream • meteorite • micrometeorite • molecular cloud • nebula • Oort cloud • parent body • planet • planetoid • radioactive** (decay, isotope) **• refractory • satellite • solar nebula • solar prominence • space probe • star • supernova**

Signed carbon

Certain meteorites are very rich in organic compounds. By comparing these substances to those in other primitive solar system objects like comets or to those in dense clouds in the interstellar medium, we hope to understand the formation of the solar nebula and determine the possible role of these compounds in the appearance of life on Earth.

CARBON-BEARING METEORITES

Organic molecules are essentially made up of carbon, hydrogen, oxygen and nitrogen. Their presence in primitive meteorites was noticed as early as 1834. In 1868, the chemist Marcellin Bertholet identified carbonaceous substances with chemical properties similar to those of terrestrial coals in the meteorite Orgueil (which fell four years earlier in the little village south of Montauban whose name it bears). The accuracy of these observations was later counteracted by the idea that these carbonaceous substances were not really part of the meteorite's original composition but instead resulted from terrestrial contamination associated with life. In 1953 the Americans Stanley Miller and Harold Urey carried out an experiment which is still famous: while trying to simulate the primitive atmosphere of the Earth (by electric discharges in hydrogen, methane, nitrogen and water), they managed to make organic products of biological interest, in particular amino acids. Today we know that this simulation of the terrestrial atmosphere was incorrect. However, this experiment showed that organic compounds (amino acids, hydrocarbons) identified in meteorites could effectively have an extraterrestrial origin, which is confirmed by the presence of compounds whose chemical composition and structure have no equivalents on our

The meteorite Orgueil, a carbonaceous chondrite, has a chemical composition very close to that measured in the solar photosphere.

Left page: comet West photographed in 1975. Comets coming from the outskirts of the solar system are rich in carbon.

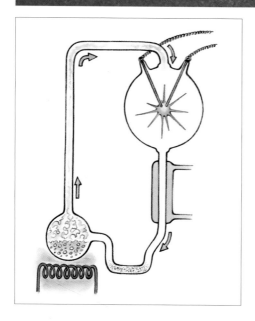

Schematic of the apparatus of Miller and Urey. A mixture of gases subjected to electric discharges simulated the early terrestrial atmosphere.

planet. Note that today nothing requires us to attribute the organic molecules identified in chondritic meteorites to life forms: they are considered as abiotic substances, that is to say not derived from life.

A GREAT DIVERSITY

The diversity of organic compounds in chondrites is impressive: apart from very varied amino acids and hydrocarbons, we find nitrogen compounds, insoluble polymers, etc. We think that hydrothermal episodes affecting carbonaceous asteroids shortly after their formation are somehow responsible for these diverse compounds. The circulation of hot water (between 40 and 80 °C) which led to the formation of clays in these bodies would have been accompanied by dissolution of some organics from which the synthesis of other compounds, such as amino acids, could have occurred.

In the world of life, amino acids are part of the make-up of proteins. Among more than 70 amino acids found in meteorites, only 8 are commonly present in proteins. Certain others are much rarer: about fifty are not present in metabolic processes. Moreover, at the end of the seventies, the origin of certain nitrogen compounds (adenine, guanine, uracil…) was the subject of heated discussions between researchers. These compounds indeed play a major biological role, because they are involved in the coding of deoxyribonucleic acid (DNA), which is the chemical support of our heredity. All of those found in meteorites are present in living matter.

SMALL AND LARGE ORGANIC MOLECULES

One general conclusion falls out from an examination of these organic building blocks: their structure is the result of a statistical recombination of simpler organic molecules, with no preferred orientation for certain chemical bonds. They were therefore formed by a series of chemical reactions, without catalysts (unlike what happens in the living world, where enzymes play a fundamental role). These "small" organic compounds represent scarcely 20% of the total organic matter in meteorites.

Most of the carbon in meteorites in meteorites is in the form of polymers (or macromolecules): giant molecules formed from a skeleton of several hundred carbon atoms to which atoms of hydrogen, oxygen, nitrogen and sulphur are attached. For lack of a suitable experimental technique, the structure of these polymers is not known in detail. Currently, neither models nor laboratory experiments explain the diversity of organic constituents in meteorites. Progress in this domain would allow us to understand the no doubt fundamental role of liquid water in the prebiotic soup.

Nucleus of Halley's comet observed by the Giotto probe in 1986.

What is a comet?

Far from the Sun, a comet is just an irregular icy nucleus, of kilometre size. When its distance from the Sun is only 4 or 5 AU, the ices vaporise due to solar heating. The nucleus then liberates gas (water vapour, carbon monoxide...) and dust formed of silicates mantled with compounds rich in carbon, oxygen and hydrogen. This material diffuses the sunlight and forms a vast luminous halo around the nucleus, the coma. This halo then stretches in the direction opposite to that of the Sun in a fine, straight "plasma tail" and a wider, curving dust tail. Comets are the remains of the small bodies of the early solar system which were thrown far from the Sun by perturbations of their orbits due to the formation of the giant planets. Today they form a vast halo (the Oort cloud) at distances up to tens of thousands of astronomical units from the Sun. The interstellar cold allows their perfect preservation. Comets which approach the Sun periodically can disappear by evaporation and dissipation of their material, or become inert after having lost all their volatile material.

INTERSTELLAR MOLECULES

The presence of small, simple molecules in space was noted as early as 1937, but it was only 35 years later that we began to understand the complexity and variety of interstellar molecules. These were detected thanks to their characteristic spectral signatures (often in the infrared or in the radio wave region). With the construction of telescopes and spectrographs allowing finer and finer analysis of light this research really took off.

The first molecules detected in the cold dense clouds of the interstellar medium (carbon dioxide and ammonia) were well known on Earth. The success of the first detections rapidly led astrophysicists to search throughout the universe for more and more complex molecules. The spectral signatures of numerous molecules were

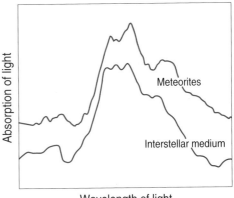

Organic molecules in meteorites have the same spectrum as certain molecules in interstellar clouds, looking at absorption of infrared light. This analysis indicates the presence of organic macromolecules in interstellar clouds. These molecules, which were present in the solar nebula, have been preserved since the formation of carbonaceous meteorites, 4.56 billion years ago.

established in the laboratory before being sought, and often detected, in space. Some of this work has a biological angle. All the same, most of the molecules which constitute the elementary building blocks of life have not been detected, so far, in the interstellar medium.

FROM THE INTERSTELLAR MEDIUM TO METEORITES

The desire to find correlations between the molecules in space and the watery mixture of organic compounds at the origin

Molecular cloud in Orion observed by the Hubble space telescope. It contains organic molecules similar to those detected in primitive meteorites.

The flybys of comet Halley

Composite photo showing the Giotto probe meeting the comet Halley.

In 1986, during the return of Halley's comet close to the Sun, no less than five space probes flew by it, three very close: two Soviet Vega probes and the European Giotto probe. An extraordinary harvest of results was obtained, modifying considerably our conception of cometary nuclei. While we expected a brilliant object, composed mostly of water ice, the Halley nucleus was the darkest object known in the solar system. This was explained by another discovery: a major fraction of the carbonaceous matter, mixed with the ice, is in the form of giant refractory macromolecules, true organic grains similar to the very black residues obtained by irradiation in the lab. They could be compounds formed before the comets themselves, in the presolar interstellar medium. This hypothesis is in accord with certain isotopic measurements. Cometary ices resemble even more the ices in cold interstellar clouds: they could be presolar remnants, whose fall to Earth would have helped add volatile constituents to the water of the early oceans, with somewhat similar deuterium/hydrogen ratios. Such a hypothesis remains to be confirmed.

Comets, sources of meteorites?

Halley's comet seen in 1986

Up until 1970, orbital theorists supposed that ordinary chondrites (75% of observed falls) must have a cometary origin, because, to their knowledge, there existed no mechanism able to transfer an asteroid fragment from a quasi-circular orbit into an elongated orbit crossing that of the Earth. Hence the idea that the most abundant meteorites were residues of cometary nuclei which had lost their volatile elements by heating during repeated passages close to the Sun. This assertion was disproved by mineralogical and petrologic observations. Moreover, radiometric ages and metamorphism also implied heat sources which would have degassed the original cometary material; in other words, it was not possible to equate ordinary chondrites and actual cometary bodies. Since then, we have discovered mechanisms of transfer from an orbit typical of an asteroid to one crossing that of the Earth, involving collisions: they are linked to gaps noted in the distribution of the asteroids as a function of their mean distance from the Sun. The very great majority of meteorites are really fragments of asteroids ejected by impacts in zones of resonance related to the attraction of Jupiter. They do not come from comets.

of life on Earth (prebiotic soup) has been a powerful motivation for the search for more and more complex organic interstellar molecules. The discovery of formaldehyde and formamide encouraged these efforts. We now know that dense clouds in the interstellar medium are veritable nurseries for organic molecules.

Nevertheless, it is the abundance of deuterium, the heavy isotope of hydrogen, which is really responsible for our making the connection between interstellar clouds and the organic compounds in meteorites. The giant organic macromolecules in meteorites have deuterium contents similar to those of the small interstellar molecules. Their precursors can therefore only be

interstellar. Chemical reactions in interstellar clouds are dominated by reactions between electrically charged molecules and neutral molecules in the gas. These are rapid reactions at the low temperatures of these clouds. They select the deuterium of the ambient hydrogen and concentrate it in the organic molecules. No other physico-chemical mechanism is conceivable today to explain the deuterium concentrations in the organic molecules in carbonaceous chondrites.

Giant organic macromolecules were also detected in the interstellar medium towards the end of the eighties, and represent one of the major forms of carbon, as in meteorites.

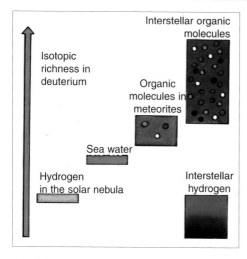

Isotopic richness in deuterium

Interstellar organic molecules

Organic molecules in meteorites

Sea water

Hydrogen in the solar nebula

Interstellar hydrogen

Isotopic abundance of deuterium (the heavy isotope of hydrogen) varies greatly in the diverse constituents of the universe. Interstellar hydrogen, the giant planets and the Sun at its formation are poor in deuterium. Interstellar and meteoritic organic molecules are rich in deuterium which shows the link between these entities. The intermediate value for the Earth (sea water) can be interpreted as the record of a mixture of these two components.

COMETS AND METEORITES

Numerous organic molecules have been detected in the gas which escapes from cometary nuclei and which feed their plasma tails when they pass close to the Sun. Analyses carried out in 1986 by the two Soviet Vega probes and the European Giotto probe during the last appearance of comet Halley indeed showed that organic matter represents two thirds of the solid particles generating the gas in the tail.

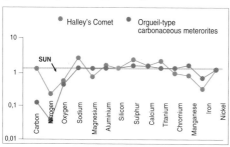

These analyses allowed us to obtain the abundances of carbon, oxygen and nitrogen, the three chemical elements essential to organic matter. The abundance of carbon is close to that found in the Sun, that is to say ten times higher than what is measured in carbonaceous chondrites. These results suggest on the one hand that these meteorites are not relics of comets, and on the other hand that the chemical composition of the solar nebula was better preserved, as far as volatile elements are concerned, in comets than in meteorites.

A RIDDLE: THE APPEARANCE OF LIFE

The terrestrial oceans could have been generated by the combined supply of carbonaceous chondrites and comets to the early Earth. Indeed, the isotopic composition of rare gases in carbonaceous chondrites, essentially contained in organic macromolecules, resembles that of the terrestrial atmosphere, but is different from what we estimate for the solar nebula. Also, the deuterium content of the terrestrial oceans is comparable to that of carbonaceous meteorites. The flux of micrometeorites, of asteroidal and cometary

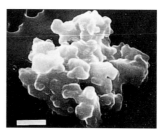

Micrometeorite observed in the electron microscope.

This figure demonstrates the great similarity in chemical composition between carbonaceous chondrites and comets. Comets seem nevertheless to have better preserved the primordial abundance of carbon and from this point of view better represent the composition of the solar nebula (the abundances are expressed as ratios with those of the Sun - horizontal line).

Stromatolites are rocks constructed by marine organisms living on Earth for four billion years.

Life on other planets

Exobiologists study the problem of life on other planets. Several approaches are possible, such as the study of the origin of life on Earth, the study of the formation and evolution of complex organic matter in the universe, the direct search for extraterrestrial living systems on the planets of the solar system, and the search for signals of extraterrestrial civilisations (SETI). Neither exploration of the solar system nor the SETI program have so far been able to provide evidence of life elsewhere. However, thanks to studies of the origin of terrestrial life, we are beginning to decipher the conditions necessary for its appearance and maintenance. Two ingredients appear indispensable for chemical evolution towards life: complex carbonaceous matter and liquid water (in the processes which allowed life to appear on Earth, about four billion years ago, meteorites and comets seem to have played a key role, by participating in the supply of carbonaceous matter). We would also like a planet with an atmosphere and a hospitable temperature (capable of maintaining water in the liquid state). In the solar system, apart from the Earth, and ignoring Europa, only the planet Mars experienced these conditions, at the beginning of its history. Structures in a Martian meteorite were indeed once interpreted as fossils, but are probably mineral in origin. If life did appear on Mars, it may have disappeared since, like the dense atmosphere and liquid water that were present there. However, there is every chance that billions of other stars in our galaxy possess a planetary system. The possibility that certain of these planets could harbour life is far from being zero.

origin, was very intense at the beginning of Earth history, and they too probably contributed organic compounds to the early oceans. The destruction of organic molecules from carbonaceous meteorites and comets in the ancient oceans could very well be the source of the first metabolisms. But for this scenario we have no geological, cosmochemical or biological evidence.

Key words: **amino acid** • **asteroid** • **chondrite** • **comet** • **contamination** • **DNA** • **enzymes** • **gap** • **hydrocarbon** • **ice** • **interstellar** • **irradiation** • **isotope** • **meteorite** • **micrometeorite** • **nebula** • **nitrogen compound** • **Oort cloud** • **organic** (compound, molecule, substance) • **plasma tail** • **polymer** • **prebiotic soup** • **protein** • **radioactive dating** • **solar nebula** • **solar photosphere** • **spectrograph**

A stormy

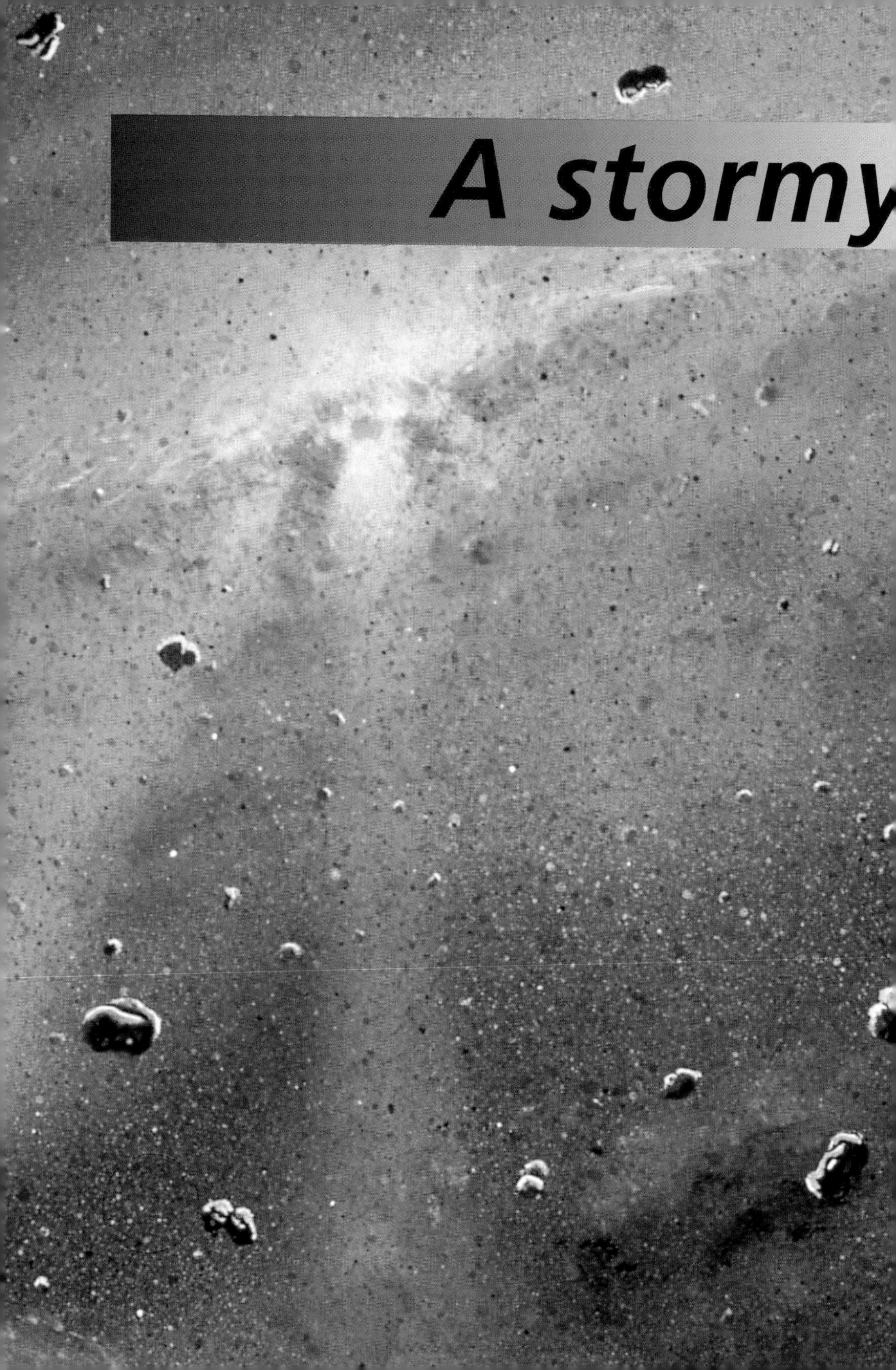

nebula

The process of planet formation still arouses many questions, but chondrules, the little spheres of melted silicate characteristic of undifferentiated meteorites, contain precious indications of the events which happened at the very beginning of the solar system.

The formation of the solar system began with the contraction of an interstellar cloud of gas and dust which was spinning on its own axis. The matter in this cloud spiralled towards the centre, where the Sun would grow, and formed a flat disk – the solar nebula. Most of this material passed in through the disk to feed the growing Sun, but the disk itself also grew by spreading outwards at the same time. We also call the nebula the "protoplanetary disk", because the fraction of its material which survived the formation of the Sun was incorporated into planetesimals and comets and then, after a sequence of accretionary events, into planets.

PRIMITIVE MATERIAL

What was the material which fed the planetesimals like and what were the temperatures in the nebula? We think that in the beginning, there was a cold cloud containing an assemblage of dust grains condensed in winds emitted by stars. However, the gas in the cloud feeding the dense protoplanetary disk must have been

Left page: artist's view of the protoplanetary disk around the young, growing Sun. The accretion of the first solid bodies has already begun.

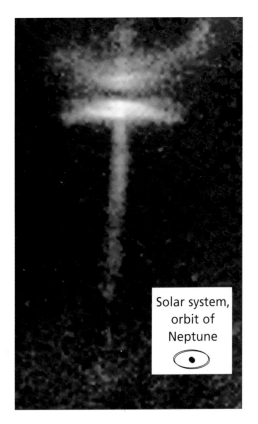

Solar system,
orbit of
Neptune

Dust disk around a star in the process of formation (Herbig-Haro 30). The disk (greenish) hides the protostar in the centre and flares towards the edges. Jets of matter (in red) are emitted by the star perpendicular to the disk (photo by Hubble Space Telescope).

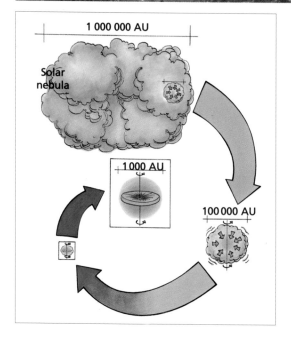

The isolation of a fragment of an interstellar cloud and its collapse on itself generate the solar nebula. Distances are expressed in astronomical units (distance from Earth to Sun). The nebula is 1000 smaller than the initial cloud and, at least near the Sun, the temperatures are high.

that most of the material used to build the planets formed by condensation in the solar nebula itself, and thus was homogenised.

A SIMPLE SCENARIO

The formation of refractory inclusions is explained well if we start with the mixture of gas and grains present in the nebula, no matter what degree of heating is invoked. If material analogous to chondrites is incompletely evaporated, the residue is enriched in refractory elements (notably calcium and aluminium). On the other hand total vaporisation followed by slow cooling of the resulting gas causes the condensation first of the refractory elements. This process could happen locally when heated material is transported turbulently from the midplane to the surface and cools by radiation to space. Refractory inclusions could thus be vaporisation residues, or the first condensates.

The nebula however must have been totally cooled down for condensation to

compressed and therefore strongly heated. In fact, temperatures varied considerably in the protoplanetary disk, from a few thousand degrees close to the growing Sun to only a few tens of degrees at the edges. Near the centre, most of the interstellar grains were heated and vaporised, while further out, in particular where comets formed, they were much better preserved. Hence in the vicinity of what is the asteroid belt today, that is to say quite far from the Sun, a very low fraction (about 1/1000) of the presolar grains was able to survive and they have recently been found in chondrites. These primitive meteorites also contain refractory inclusions in which some minerals contain elements with an isotopic composition different from that of solar material: these minerals were formed in the solar system from grains produced earlier in distinct stellar environments. However, despite these interstellar traces, we think

Perspective view of the protoplanetary disk. Dust and gas rotate around the young Sun before agglomerating to form planetesimals.

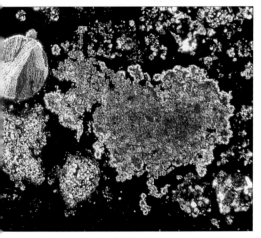

Refractory inclusion in the Allende meteorite. The "fluffy" nature of the inclusion suggests that it formed directly from condensation (at high temperature) within the nebular gas.nébulaire.

Another refractory inclusion in the Allende meteorite. The texture indicates that it was melted.

The condensation sequence

Just as water vapour condenses to frost, the gas of the solar nebula could have condensed to mineral grains, or even liquid droplets if the pressure were high enough. In a nebula, where the pressure is very low (about a hundred thousandth of terrestrial atmospheric pressure), minerals condense in the opposite sequence to the one they follow when they evaporate. At high temperature, the minerals formed first, or surviving intense evaporation, are oxides and silicates of aluminium and calcium, which we find in refractory inclusions. After the condensation of refractory minerals come the main minerals of chondrites: iron and silicates rich in magnesium. These minerals later react with the ambient gas, as cooling continues, and more volatile elements are little by little incorporated into the solids: thus, for example, most of the iron was probably converted into iron sulphide. Finally, at the lowest temperatures, the condensation of organic compounds and ices takes place.

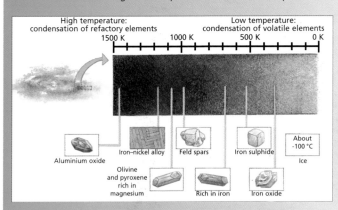

High temperature: condensation of refractory elements

Low temperature: condensation of volatile elements

1500 K   1000 K   500 K   0 K

Aluminium oxide

Iron–nickel alloy

Feld spars

Iron sulphide

About -100 °C

Ice

Olivine and pyroxene rich in magnesium

Rich in iron

Iron oxide

As the nebula cooled (over time and as a function of distance from the Sun), new minerals condensed, from the most refractory (formed at high temperature) to those including the most volatile elements (organic compounds and water ice).

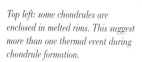

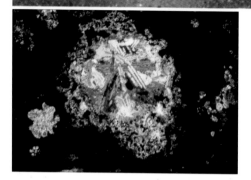

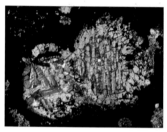

Top left: some chondrules are enclosed in melted rims. This suggest more than one thermal event during chondrule formation.

Top centre: we frequently find crystals in chondrules different from all the other crystals and which have manifestly survived (at least in part) the melting episode. Are they inherited from earlier generations of chondrules?

Top right: the existence of "compound" chondrules is considered an indicator of successive generations of chondrules, because one must be still plastic and the other already solid for them to have stuck together.

occur on a large scale. According to the simplest scenario, the asteroids and planets would be formed by accumulation of grains condensed during this cooling. However, natural processes are rarely simple. The nebula was gigantic, stirred by violent storms, and its history was complex.

CLUES FROM CHONDRULES

Chondrites contain evidence of transient heating events. The particles which make them up (chondrules and also some refractory inclusions) result from melting in the solar nebula, under microgravity conditions. Now as astronauts regularly see, under such conditions a liquid takes the form of a sphere. This is what chondrules did before they froze during cooling. Crystallisation experiments on droplets of this type show that cooling for minutes to hours is sufficient to produce crystals identical to those they contain. The whole nebula could not have cooled down in such a short time. This implies that chondrules were formed by transient, local events. In addition, certain observations suggest that chondrules could have formed several million years after refractory inclusions. In this case, the latter could have formed by different processes.

Captain Haddock has no luck with his whisky! As Hergé reminds us, the form taken by a liquid under microgravity conditions is that of a sphere. This is how the round form of chondrules, melted in the solar nebula, is explained. (© Hergé – Moulinsart – 1966)

What is a chondrule?

Chondrules are small spherules of silicate material that experienced melting before incorporation into chondritic meteorite parent bodies. Their abundance in chondrites and hence in many asteroids implies that melting of small particles was a common phenomenon in the early solar system. Chondrules are silicate particles (containing olivine, pyroxene and glass, and minor iron-nickel metal and iron sulphide) in chondrites. They are thought to have been melted, because of the common droplet form and their texture with euhedral to skeletal (rapidly grown) crystals set in residual glass.

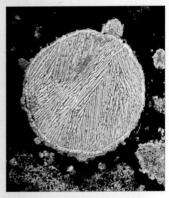

Barred olivine chondrule in the Allende meteorite. It consists of a few parallel-plate crystals of olivine, with glass between the plates, indicating that the chondrule was very thoroughly melted.

Chondrule with porphyritic texture in the Ornans meteorite. Laboratory experiments show that it takes a few hours to grow such crystals from a silicate melt.

CHONDRULE FORMATION

A large number of chondrule formation mechanisms have been proposed, from asteroid collisions to lightning. The true process remains uncertain.

Assuming that the Sun behaved like young stars of the same mass observed today, the accretion of material from the interstellar cloud to the disk and the growing Sun was not steady. It was punctuated by episodic events, marked by sudden increase in luminosity associated with a major increase in accretion rate and a considerable temperature rise close to the Sun. It is also possible that late-arriving gas-dust clumps generated shock waves in the disk causing more local temperature rise. Finally collisions of grains due to turbulence in the nebula could have led to a separation of electrical charges leading to gigantic lightning bolts. However, lightning, being a smaller scale mechanism, would probably have lead to major temperature differences locally. Nowadays, we think that shock waves are the most likely heat source for chondrule formation as they seem to have been melted in a relatively uniform manner, with few unmelted aggregates or extensively evaporated spherules.

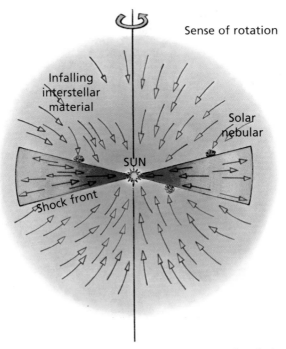

Infalling
interstellar
material

Sense of rotation

Solar
nebular

SUN

Shock front

Material from the parent interstellar cloud gradually falls into the disk, the inner part of which feeds the Sun as the outer part spreads outwards. The infall may be continuous, but also in the form of clumps which generate shock waves. This is one possible way of generating high temperatures in the disk so as to form chondrules.

FROM THE NEBULA
TO PLANETS

It is difficult to understand the very first stages of accretion, which allowed grains in the solar nebula to accumulate to form planetesimals. In a turbulent, hot nebula stirred by thermal convection, dust grains and fluffy aggregates are more likely to have been carried by the winds than to have settled and concentrated in the midplane of the protoplanetary disk. Calculations show that particles of the size and density of chondrules could have been concentrated in stagnant zones between eddies, leading to clumps of kilometre size. The latter could have later slid en masse to the midplane of the disk, allowing accretion to start. The process must have been quick and sudden, as volatile elements concentrated in gas and dust after chondrule formation have, at least in part, been incorporated into asteroids instead of being dissipated with the nebular hydrogen.

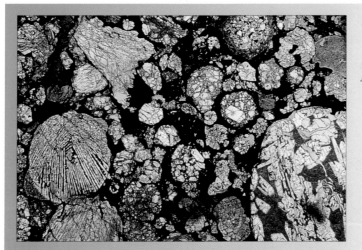

Some chondrites (ordinary chondrites) can contain up to 80% chondrules. The chondrules are enclosed in a fine opaque matrix (Semarkona meteorite x 13)

The formation of chondrites

Some chondrules are rich in refractory elements and poor in volatiles. The volatiles could have been missing from the chondrule precursor material or lost during the melting process. In some cases, the volatiles seem to be concentrated in the matrix of fine dust cementing the chondrules into the rock. Chondrites must have accreted relatively rapidly, to retain the refractory material and most of the volatile fraction in almost solar (bulk solar system) proportions, without separation of the chondrules and fine dust. One possible explanation is that most of the fine dust rode down to the midplane as rims grown on chondrules, and the chondrule matrix probably originated mostly as fluffy rim material knocked off as the chondrules were incorporated into their parent bodies.

Key words: **accretion • aggregate • asteroid • astronomical unit • chondrule • chondrite • comet • condensation • convection • crystal • crystallisation • glass • interstellar** (cloud, grain) **• isotopic composition • matrix • meteorite • nebula • nebular gas • parent body • planet • planetesimal • presolar grain • protoplanetary disk • refractory** (element, inclusion) **• shock wave • silicate • star • stellar wind • vaporisation • volatile**

The age o

he solar system

We have known the age of the solar system for over forty years and more recently its date of birth was defined precisely: 4,566 million years ago. The Earth formed over the next hundred million years and life began to develop about 4 billion years ago.

Despite this very respectable age, the Sun and its planetary cohort appear young in comparison to the observable Universe, whose formation goes back about 15 billion years. Several billion years elapsed after the formation of our galaxy before a fragment of a molecular cloud in one of its spiral arms gave rise, by contracting, to an infant Sun and a swarm of planetesimals. Finally, after a history of accretion, some of these planetesimals became the planets of the solar system...

CHOOSE A CLOCK

The age of the solar system and the history of its formation are determined with the help of a battery of clocks, which count time in meteorites. These clocks are nuclear in character. Some atomic nuclei are radioactive: they are unstable and spontaneously transform into other nuclei. We speak of the decay of the "parent" nucleus into the "daughter" nucleus.

The disintegration of a given nucleus is instantaneous, but it may happen after a

Left page: thin section of the achondrite Angra dos Reis (crossed polarisers). This stone is one of the oldest differentiated materials in the solar system (4.558 Ma).

relatively long time. The "half-life" is the interval of time at the end of which half the parent nuclei initially present have disintegrated. Depending on the element, it can vary from a fraction of a second up to several tens of billions of years. Each clock ticks with its own rhythm, governed by this half-life, and it is possible to determine the timing of a phenomenon using the most suitable clock. Thus, to characterise the initial history of the solar system, we have two kinds of clocks: those which work over a long period, to measure ages of several billion years, and those which stop at the end of a short period, to measure intervals of one to two hundred million years. In this field, we use the abbreviations Ga (giga

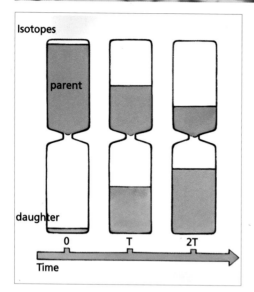

Isotopes

parent

daughter

0   T   2T

Time

Left; at the end of one "half-life" (here T) specific to the radioactive element, half the unstable "parent" atoms are transformed to "daughter" atoms. The budget of parent and daughter atoms in a rock constitutes a radioactive clock.

of chondrite contained sixty seven thousand billion uranium atoms. This dates the switching on of the clock relative to today.

While dating with long-period clocks is always done following this principle, the ages in fact depend on the chemical history of the material. More exactly the results define the time since which the parent and daughter atoms in the sample behaved as a closed, isolated system, which neither gained nor lost material due to external influences. This could correspond to the

years) or Ma (mega years) to indicate billions or millions of years.

READING A LONG-PERIOD CLOCK

A one-gram fragment of chondrite contains an infinitesimal amount of uranium, about 10 billionths of a gram. This tiny mass represents a gigantic number of atoms, twenty five thousand billion. This number was even greater when the meteorite was formed because a fraction of the initial uranium atoms have decayed giving lead atoms.

By counting the radiogenic lead atoms and the remaining uranium atoms in this fragment, the time elapsed since the uranium began to decay in the rock can be calculated. The counting is done with a mass spectrometer, after extraction and chemical separation of the traces of uranium and lead. The work is carried out taking care not to contaminate the sample with terrestrial uranium and lead.

The "uranium-lead" age thus obtained is close to 4.5 Ga. At that time, the fragment

Examples of radioactive transformations

Radioactivities of long period

Transformation	Half-life (in billions of years)
rubidium-87 / strontium-87	48.8
rhenium-187 / osmium-187	43.5
uranium-238 / lead-206	4.468
potassium-40 / argon-40	1.25
uranium-235 / lead-207	0.704

Radioactivities of short period

Transformation	Half-life (in millions of years)
iodine-129 / xenon-129	16
palladium-107 / silver-107	6.5
aluminium-26 / magnesium-26	0.7
calcium-41 / potassium-41	0.1

formation of the rock, but a simple heating can perturb the system. For example, while the parent isotope may be securely in place in the crystal lattice, the daughter is generally less secure because it does not have the same chemical properties as the parent. It may then try to migrate if that is possible; if it succeeds, the parent-daughter system is no longer closed. In this case, the clock will no longer indicate the time elapsed since the formation of the rock, but either the age of the last event which set the stop-watch back to zero, or a hybrid age with no precise meaning.

FAMILY PORTRAIT

In the "unshocked" chondrites, long-period clocks indicate very similar ages between 4.56 and 4.48 Ga. They were started almost at the same time and have ticked away without perturbation right up to today.

This age identifies chondrites as among the oldest materials known in the solar system. The oldest terrestrial rocks are dated at 3.8 Ga; the oldest lunar rocks at 4.4 Ga.

In addition, during mineralogical transformations within their parent bodies (metamorphism), the chemical composition of chondrites has only been slightly

Principal of radioactive dating methods

Radioactivity results from the instability of the atomic nucleus. It is independent of the chemical and thermodynamic environment of the isotopes (atomic nuclei characterised by the number of protons and neutrons that they contain). That is why it acts as a clock that works in an identical manner no matter what mineral, rock or planet is studied. The half-life is the time necessary for the initial abundance of atoms of the parent isotope to decrease by half in transforming to the daughter isotope. This period is the signature of each radioactivity.

Radioactive dating is based on the fact that the parent and daughter isotopes belong to different chemical elements and on their presence in geological materials. The radioactive clock is started at zero time by the thermal event that modifies the abundance of parent and daughter isotopes: transformation of minerals in a rock, differentiation of a planet or cooling after the accretion of a planetary body.

Study of the formation of the solar system and planetary objects uses short-lived isotopes (half-lives 1 to 100 million years), which are today "extinct".

A long-lived isotope defines an absolute age if the decay happens in a closed system. We can date to less than about a million years the most ancient meteorites using the decay of uranium-238 and uranium-235, which transform respectively to lead-206 and lead-207 with very different half-lives.

The Earth began to form 4.56 billion years ago, but its formation took about 120 million years.

The age of the Earth

The age of the Earth becomes a subject of scientific debate in the 19th century. The geologist Lyell and the biologist Darwin propose an Earth aged at least several hundred million years. The physicist Thomson (Lord Kelvin) argues for an age of a few tens of millions of years by calculating the time needed for the cooling of our planet from a melted state. Rutherford closes the debate in 1929: by measuring the quantity of helium produced in rocks by the decay of uranium, he defines a radiometric age of about a billion years. A new confrontation then appears between astronomers and geologists. The American astronomer Hubble calculates an age for the Universe of about 2 Ga. The response comes from C. Patterson in 1953-1955: by measuring the isotopic composition of lead, he shows that the Earth and meteorites have the same age, 4.55 Ga; this is also the age of the solar system.

Today, the age of the solar system is known precisely as 4.566 Ga. Our planet then developed by accretion of planetesimals over the next hundred million years that followed. It is also at this time that the great terrestrial sub-units, core, mantle and primitive atmosphere, differentiated.

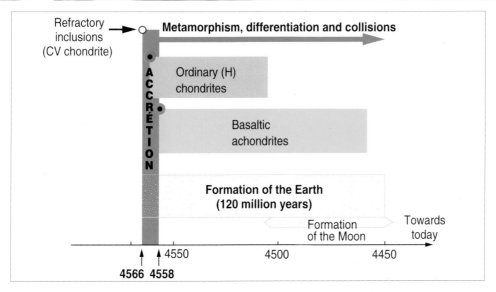

Refractory inclusions, the first solar system objects, were formed 4.566 billion years ago. Formation of the chondrites which contain them and of achondrites lasted about 8 million years. Heating in the parent bodies was able to modify mineral structures for another 100 million years. The Moon could have formed after the collision of a giant planetesimal with the Earth, about 4.5 billion years ago. The oldest lunar rocks are dated at 4.4 billion years.

changed. The formation age of chondrites then corresponds essentially to the formation age of their parent bodies. The mineralogical transformations took place within an interval of 0.15 Ga, between 4.56 and 4.50 Ga. This period reflects the weak thermal activity, which affected planetesimals soon after their accretion.

Dating of differentiated meteorites, irons and basaltic achondrites, defines the time when their parent bodies cooled after melting produced metal cores surrounded by silicate. The measurements yield a range of ages, from 4.56 to 4.45 Ga, similar to that defined by chondrites. The volcanism that occurred on these bodies is very old: the lavas poured out on their surfaces, represented by eucrites, show it began between 4.56 and 4.53 Ga. Thus thermal activity in the parent bodies of differentiated meteorites was contemporaneous with that which affected chondrite parent bodies in a less intense manner.

Long period clocks used for shocked meteorites generally define varied ages, usually younger than 4.5 Ga. They have been disturbed since they were originally started at 4.56 Ga. Some may have been reset to zero by impacts. This is the case for the potassium-argon (K-Ar) clock: the argon gas generated by the decay of potassium has a tendency to diffuse out of material heated by shock. The K-Ar age will then date the collision, which affected the parent body.

STOPPED CLOCKS

The formation of the solar system started with the formation of the solar nebula by contraction. Before this, the parent cloud of gas and dust continually took in products ejected by stars. However, contraction partially isolated the solar nebula from the interstellar medium. The presolar material was no longer fed by radioactive elements

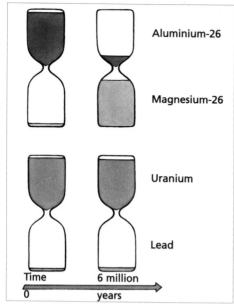

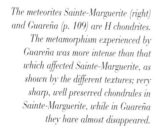

Short period (above) and long period (below) clocks. All the aluminium-26 will be transformed to magnesium-26 in 7 million years, while 37% of the uranium will remain after 4.5 billion years.

(from stellar explosions), except by shock waves. Parent bodies formed "shortly" after the isolation of the solar nebula were able to incorporate short period radioactive elements while they were still active. They thus decayed within this material.

In the latter case, there are today no longer any parent isotopes: they all decayed 4.56 Ga ago. However, we know how to identify and count the daughter atoms, which resulted from the disintegration of the parent atoms in this material. The quantity of daughter isotope indicates the time when the host material began to trap it, but it is not possible to place this moment on an absolute time scale, as in the case of long period clocks. On the other hand, comparisons between different classes of meteorites are possible if they initially contained the same amount of parent isotope. This amount can be evaluated by considering the processes of synthesis of chemical elements within stars likely to have exploded just before. In addition, an estimate of the time interval between the isolation of the solar nebula and the formation or cooling of planetary materials is possible: it depends on the quantity of parent isotope injected into the solar nebula.

A MAJOR DISCOVERY

The first discovery of the decay of a short period radioactive element, iodine-129, in meteorites, happened in 1961. Since then, xenon-129 derived from iodine-129 has been found in most meteorites. This discovery indicated that all the meteorite parent bodies were formed in a relatively short interval of time, at most 20 million

The meteorites Sainte-Marguerite (right) and Guareña (p. 109) are H chondrites. The metamorphism experienced by Guareña was more intense than that which affected Sainte-Marguerite, as shown by the different textures; very sharp, well preserved chondrules in Sainte-Marguerite, while in Guareña they have almost disappeared.

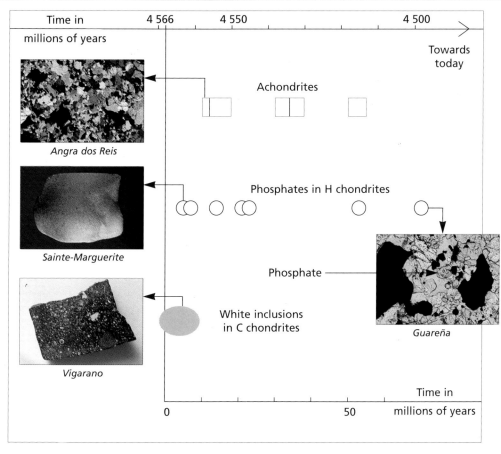

Refractory inclusions, the first objects in the solar system, define the birth date of the solar system (time 0). The lead-lead chronometer allows us to specify very precisely the ages of other solar system objects. In particular, phosphates in the most metamorphosed (type 6) Guareña are "younger" than those of the type 4 (Sainte-Marguerite), as reheating has mobilised the atoms in the crystal lattices for longer.

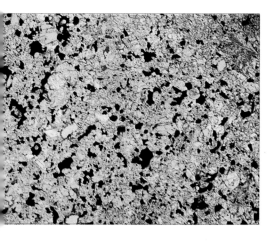

years, which is compatible with estimates derived from long period radioactive transformations. Thus the age of the solar system was bracketed: it is between 4.55 and 4.57 billion years. In addition, this result showed that less than 150 million years elapsed between the last stellar event that generated iodine-129 and the formation of parent bodies.

Since then, several other radioactivities with even shorter periods have been demonstrated: they help to identify the astrophysical context that spawned the solar system and to define its early history.

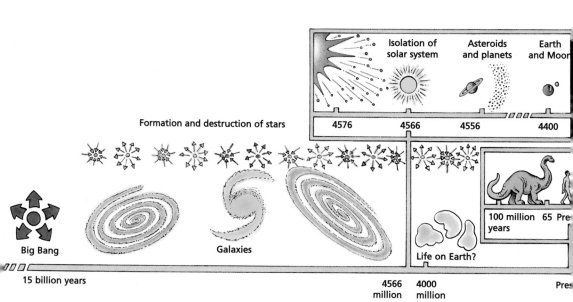

The solar system formed from hydrogen and helium generated by the Big Bang and star dust formed later. The last star that contributed material exploded only about a million years before the isolation of the nebula. Planets and asteroids began to appear a few million years later. The origin of the Moon is still a hypothesis, but the oldest lunar rocks date from 4.4 billion years.

BIRTH DATE

The birth of the solar system was rapid. We are able today to specify the date of this event by a combined reading of long period and short period clocks in material formed from the solar nebula. The most primitive carbonaceous chondrites contain refractory inclusions, which were formed at high temperatures in the solar nebula. In 1976, the presence of aluminium-26, which has a very short half-life (0.7 million years) was detected from the excess of magnesium-26, the daughter element, in the feldspar of these inclusions. Aluminium-26 was a very efficient heat source for heating planetesimals. Because of its short period, it must have been produced at most three million years before its incorporation in the feldspar. This period of time was reduced to

less than a million years with the discovery, in 1995, of excess potassium-41, corresponding to the radioactive decay of calcium-41 (with a half-life of 105,000 years).

Moreover, dating the refractory material in carbonaceous chondrites with the uranium-lead clock defines a very precise age: 4,566 plus or minus 2 million years. The combination of these two chronometric data gives us the birth date of the solar system – 4,566 million years ago. The level of uncertainty on this number does not exceed three million years and with the progress of isotopic research, we can be sure that it will become even less ...

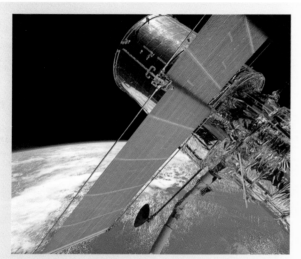

The Hubble Space Telescope, named in honour of the celebrated astronomer, is responsible for a considerable advance in knowledge in the field of science of the Universe.

The age of the Universe

In the twenties, Hubble establishes the existence of galaxies outside the one that contains the solar system. Most show a light spectrum shifted towards the red, which indicates that they are moving away. In 1929, Hubble discovers that they move away at speeds proportional to their distances. The simplest explanation is that the Universe is expanding. This phenomenon is regarded as the consequence of an initial explosion, the Big Bang.

Originally incredibly dense and hot, the Universe has cooled since by expanding. Three methods allow us to estimate the time that has elapsed since the Big Bang, which we call the age of the Universe:

- estimation of the age of the oldest stars in our galaxy, based on the understanding of stellar evolution acquired in astrophysics. The age obtained is between 14 and 17 billion years.

- radioactive dating of chemical elements synthesised in stars in our galaxy, based notably on measurements of the abundance of certain isotopes in meteorites. This leads to the assignment of an age of 13 to 19 billion years to our galaxy.

- measurement of the rhythm of expansion of the Universe, based on observations on galaxies. The values obtained in the course of the last two decades vary between 50 and 100 km/s per million parsecs (the parsec is a unit of distance used in astronomy, equivalent to 3.26 light years). With the most recent values, near 80 km/s we calculate that the age of the Universe must be between 8 and 12 billion years, which is in conflict with the results obtained by the two other methods.

Key words: **accretion • achondrite • asteroid • Big Bang • chondrite • chondrule • core of the Earth • contamination • crystal lattice • dating • differentiation • disintegration • eucrite • galaxy • half-life • interstellar medium • iron • isotope** (parent, daughter) **• light year • mantle of the Earth • mass spectrometer • metamorphism • meteorite • molecular cloud • nucleus • parent body • parsec • planet • planetesimal • radioactive** (chronometer, clock, dating, decay, element, nucleus) **• radioactivity • radiogenic • radiometric age • refractory inclusion • shock wave • silicate • solar nebula**

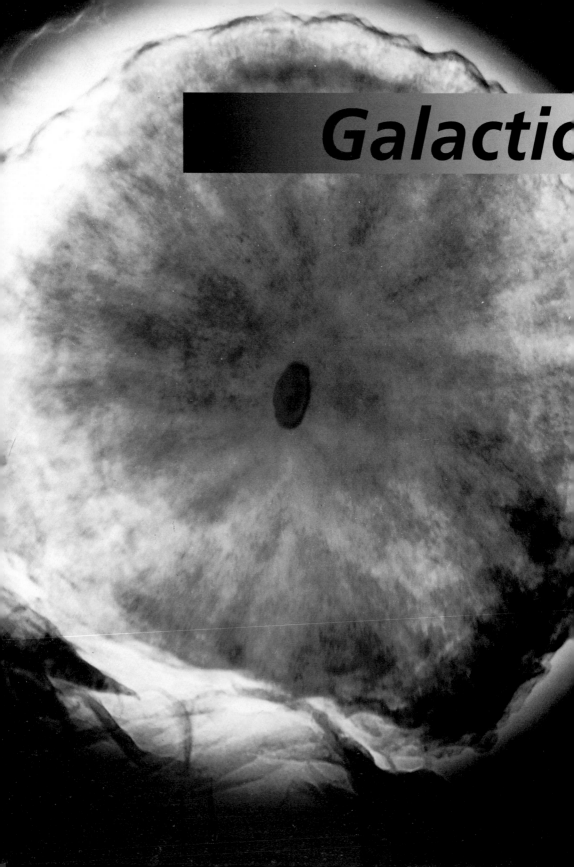

Galactic

ossils

Thorough analysis of meteorites finally allows us to study stardust in the laboratory. We suspected its presence for a long time without being able to locate it.

WHEN ISOTOPES TALK...

Until 1970, with a few rare exceptions, no isotopic differences between lunar, terrestrial and meteoritic samples were detected. This apparent similarity in the material of different solar system bodies supported the model, now obsolete, of an initially hot solar nebula, where material was vaporised and well mixed, thus losing every trace of its origin. Mass fractionations, like those caused by evaporation and condensation in the early solar system, explained the small variations in certain isotopic ratios demonstrated in meteorites. However, to locate the source materials of the nebula itself, it was necessary to detect abnormal isotopic abundances (or isotopic anomalies), the products of processes occurring within stars, the giant factories for making the chemical elements. A few examples were known, notably of two noble gasses, xenon and neon, but the carrier

Left page: this presolar grain (seen in a transmssion electron microscope) consists of well crystallised graphite grown on a grain of titanium carbide (at the centre). The ability of these grains to enclose others tells us about the physical and chemical properties of the stellar gas in which the grains condensed.

minerals of these gasses remained to be discovered. Finally, in 1973, in the Allende meteorite, scientists identified refractory inclusions rich in anomalous isotopic ratios; ratios close, moreover, to those that certain astrophysical models predict for stars.

THE CHEMISTRY OF STARS

Most astrophysicists today consider that the Universe, such as we find it, began by a gigantic explosion, the Big Bang. In this initial furnace the lightest chemical elements, hydrogen, helium, and one isotope of lithium, lithium-7, were produced. All the other elements were produced progressively, later in stars by what we call stellar nucleosynthesis. From the primordial hydrogen and helium, the first generation of stars produced elements such as carbon and oxygen. At the end of their lives, stars, in particular the most massive ones, eject gas and dust into the interstellar medium, and this material later serves to form new stars. A large number of generations of stars must have succeeded one another before the formation of the Sun, 4.566 billion years ago.

The matter ejected into the interstellar medium by dying stars, red giants and

Isotopic anomalies

Every deviation from the average isotopic composition of the solar system is an "isotopic anomaly". The most spectacular anomalies are found in surviving presolar grains. However, solids formed in the early solar system can also have significant anomalies, if they are generated from presolar material that was not completely homogenised. Isotopic anomalies can result from the decay of short-lived isotopes in a grain that formed early in the solar system. More generally they are created by stars, which are nuclear reactors.

The isotopic composition of the solar system results from the thorough mixing of numerous components derived from diverse stellar sources. Thus the isotope oxygen-16 to oxygen-18 ratio of the solar system is close to 500, but that of presolar grains varies from 2.7 to 40,000. The stars that fed the solar nebula therefore produced at least this range of oxygen.

Interpreting isotopic anomalies is not always easy. For example, certain refractory inclusions in primitive meteorites contain an excess of oxygen–16 that could equally well be an indicator of a process in a star older than the Sun as the result of a chemical reaction in the solar nebula.

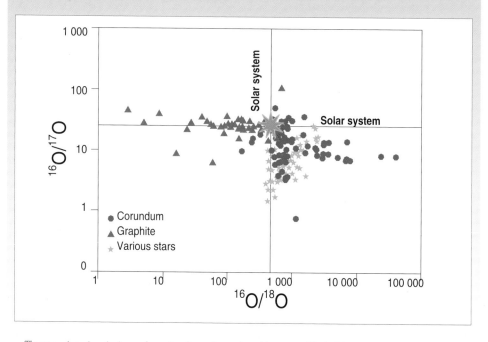

The oxygen isotopic ratios in presolar grains of corundum and graphite measured in the laboratory are more diverse than those measured by spectroscope in the atmospheres of diverse stars. The grains therefore yield information which cannot be obtained by astronomy.

supernovae, condenses into solid grains which trap the specific local isotopic composition. This is how we can identify stardust in meteorites: its anomalous isotopic signature is the sole criterion, which allows us to recognise it. This stardust has survived the processes of destruction and regrowth operating during the formation of the solar system.

extracted from the Allende meteorite. The refractory inclusions in this carbonaceous chondrite contain large isotopic anomalies in calcium and titanium, elements produced close to the centres of massive stars, as well as in heavier elements, such as barium, synthesised by the capture of neutrons in material ejected by supernovae (r process).

Later hibonite, a very refractory mineral, was isolated in the Murchison meteorite. It turned out that its bluish crystals contain an

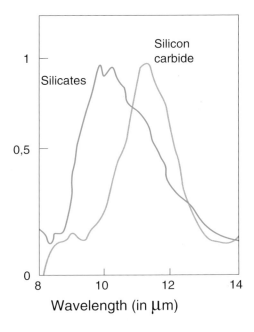

Infrared emission spectra indicate the presence of dust envelopes around red giants. Stars rich in carbon generally show an emission line at 11.15 mm, characteristic of silicon carbide, whereas stars rich in oxygen show a line at 9.8 mm, the signature of silicates.

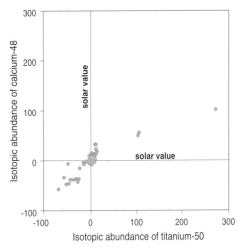

Hibonite, a refractory mineral extracted from carbonaceous chondrites, shows either large excesses or large deficits in the neutron-rich isotopes of calcium and titanium.

FIRST INDICATIONS

The first sample of which we can say that it contained presolar traces was

excess or deficiency of calcium-48 and titanium-50. The combined presence or absence of these two neutron-rich isotopes are signatures of two different types of supernova. Distinct grains coming from these two types of supernova must then have been present in the early solar system and incorporated in these hibonites. However, refractory inclusions and hibonite are not intact circumstellar condensates: they only incorporated isotopically anomalous stellar components at the moment of their formation in the early solar system. The isotopic analysis of the resulting material informs us

The origin of the elements

Most of the chemical elements are made in stars, by nuclear reactions, starting from hydrogen. A star has temperatures (from a million to a billion degrees) and particle concentrations favourable to nucleosynthesis. The first reactions lead to the fusion of hydrogen into helium, then helium into carbon. Low mass stars do not go beyond this stage. Supplementary additions of helium nuclei create oxygen, neon and a whole collection of intermediate elements in medium mass stars. In the most massive stars (more than 9 solar masses), nucleosynthesis will produce the elements up to iron. As the latter has the most stable nucleus nuclear fusion stops at this point. The heavier elements are produced either by rapid capture of neutrons in a high neutron density environment ("r" process, which takes place in supernovae), or by slow capture at low density ("s" process, which takes place in red giants).

hydrogen

helium

carbon, oxygen, neon...

silicon, sulphur (silicon-28, titanium-44)

nickel, iron (titanium-44)

Massive stars, in which nuclear reactions go as far as producing iron, end up resembling an onion. Products of nuclear burning reside in different shells, with more evolved products towards the core.

about the physico-chemical conditions prevalent at the very beginning of the solar system, as well as about the processes of nucleosynthesis, which preceded its formation.

STARDUST

Finally, in 1987, the first true presolar grains were isolated, after a twenty-year hunt to identify the carrier minerals of the isotopically anomalous rare gasses. These minerals represent less than 1/1000 of the matrix of carbonaceous chondrites but, by chance, they are extremely tough. We can isolate them after total dissolution of the matrix containing them with very aggressive acids. This stardust condensed in stellar winds from red giants and in the gas ejected by stellar explosions. Generally, all the chemical elements present in these grains have isotopic compositions, which reflect the stellar environments they come from, and which are completely different from those of solar system material. These grains

How do stars evolve?

All stars are born from pre-existing dust and gas, and die either in fantastic explosions or by cooling slowly when their internal nuclear fire goes out. Massive stars burn their nuclear fuel much quicker than low mass stars, because of higher internal pressures and temperatures, disappearing in a few million years.

In contrast, the Sun is already almost 4.6 billion years old, and will live about as much again. The mass of a star controls its lifetime. When the hydrogen of a

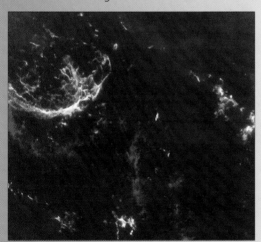

The Hubble Space Telescope has seen relics of a supernova explosion in a neighbouring galaxy, the Magellan clouds. Oxygen-rich material (blue-green filaments) ejected by the explosion at great speed is mixed into the local interstellar medium.

medium mass star is used up, the outer part expands, transforming the star into a red giant. In the case of a massive star, fuel is exhausted after an iron core is produced and contraction starts. Formation of a hard neutron core stops the collapse, and in a rebound an enormous explosion is produced, ejecting the outer layers of the star: it becomes a supernova. During the explosion, a shock wave passes through and heats ejected material, resulting in so-called "explosive" nucleosynthesis. During red giant or supernova stages, the synthesised products are ejected into the interstellar medium and grains condense. They constitute the seeds of other stars.

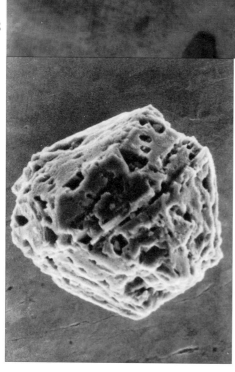

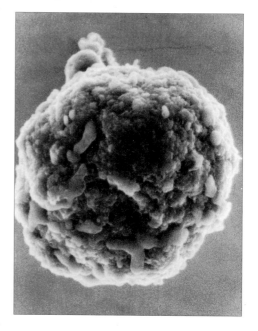

Electron microscope image of presolar grains extracted from Murchison (each measures about 5 μm). The graphite grain (below) was probably formed in the carbon-rich envelope of a supernova, a massive star which dies in a gigantic explosion. The silicon carbide grain (left) probably comes from a red giant rich in carbon.

are samples of stars and can be analysed in detail in the laboratory. Their isotopic and chemical compositions teach us about the processes of synthesis of atomic nuclei that take place in different types of star so as to produce chemical elements with distinct isotopic ratios.

RED GIANTS AND SUPERNOVAE

Three types of presolar grains rich in carbon, diamond, silicon carbide and graphite, have been discovered thanks to the anomalous rare gasses they contain. These crystals cannot form in a medium rich in oxygen like the solar system. Two other types, corundum (aluminium oxide) and silicon nitride, have been identified by isotopic analysis on the ion probe. With the exception of diamond, all these grains have sufficient size to be individually analysed with this instrument. Their isotopic compositions, compared to nucleosynthetic theories, reveal that these grains come principally from two kinds of stars: red giants, stars of relatively small mass at an advanced state of evolution, which swell and loose a large amount of material; and massive stars which explode as supernovae at the end of their short lives.

The corundum grains and the majority of the silicon carbide grains seem to come from red giants. The range of isotopic ratios

Isotopic analysis

Measurements of isotopic ratios can in principle be made using any property which depends on atomic mass. For example, spectroscopic measurements of isotopes in stars are possible because the frequency of molecular vibrations depends on the exact mass of the atoms and the optical spectra emitted by cool stars show distinct lines for different isotopes.

In the laboratory, isotopes are counted in a mass spectrometer. The "ion probe" is a mass spectrometer that does not require chemical separations before the sample is analysed. The minerals can therefore be directly analysed in situ in a rock, which allows us to choose the best spots, and make comparisons from one point to another in the rock. In addition, the ion probe is capable of analysing sub-micron samples, and thus individual presolar grains.

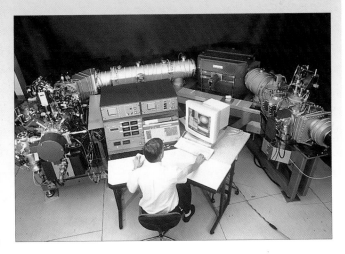

Opposite: an ion microprobe.

of oxygen, silicon and titanium that we observe indicate that a great number of different stars must have fed the solar system with dust grains.

Most of the graphite grains, 1% of the silicon carbide grains and all the silicon nitride grains apparently come from supernovae. Indeed, they do contain excesses, compared to characteristic abundances in the solar system, of oxygen-18 and, especially, silicon-28 and calcium-44, the latter being produced by the radioactive decay of titanium-44. Silicon-28 and titanium-44 are only produced in supernovae, near the centres of the stars. However, the grains rich in carbon and nitrogen must be formed in the outer shells, where helium still exists. The simultaneous presence of all these isotopic anomalies in the grains proves that, during the explosion of a supernova, the material ejected by the star experiences enormous turbulent mixing. Astronomers have indeed observed this phenomenon.

This molecular cloud in the M 16 nebula is a stellar nursery. It is illuminated by nearby hot stars, already formed out of local material.

The interstellar medium

The interstellar medium contains low concentrations of gas and dust heterogeneously distributed, from almost empty regions to dense molecular clouds. New stars are born in these clouds, when a dense region undergoes gravitational collapse.

The chemical elements are distributed in a heterogeneous fashion in this medium. Aluminium and silicon are absent from the gas but trapped in dust grains.

At the extremely low temperatures of these clouds, chemical reactions can still take place, and lead to unusual isotopic ratios for hydrogen and nitrogen, if the molecules are ionised. Ionisation is guaranteed because of cosmic rays, very energetic particles probably accelerated in supernovae, which traverse the interstellar medium. In the clouds, we detect large excesses of deuterium which correspond to what has been measured in primitive meteorites, Halley's comet and certain interstellar dust grains, which shows that the early solar system has incorporated material from molecular clouds.

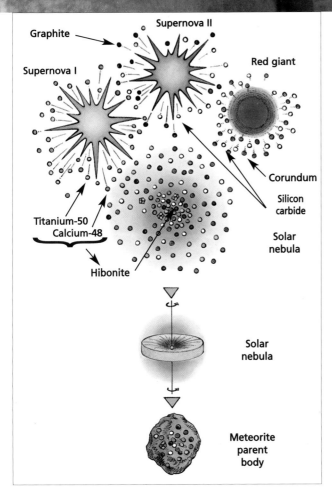

Graphite

Supernova II

Supernova I

Red giant

Corundum

Silicon carbide

Titanium-50
Calcium-48

Solar nebula

Hibonite

Solar nebula

Meteorite parent body

Supernovae and red giants which came before the formation of the solar system seeded the interstellar medium with presolar grains. Grains of hibonite showing anomalies in calcium-48 and titanium-50 formed within the solar nebula derived from this mixture.

A CLOCK THAT HAS STOPPED

The silicon carbide grains are rich in magnesium-26, formed by radioactive decay of aluminium-26. The initial quantity of aluminium-26 in these grains was much greater than that in the refractory inclusions in the Allende meteorite: the highly variable aluminium-26/aluminium-27 ratio can attain values close to unity, while in the refractory inclusions it is only on the order of 1/10,000! However, the aluminium-26 was totally transformed to magnesium-26 and the chronometer was thus extinct well before the arrival of the grains in the solar nebula. Although aluminium-26 could have been a heat source for asteroidal parent bodies, and can be used to construct a fine time scale of the first moments in the formation of the solar system by measuring sufficiently primitive materials, the effects due to decay of this isotope in the silicon carbide are fossil.

Isotopic analysis is an unusual way of thinking in the laboratory about the interior of stars. Nucleosynthetic theory has provided the tools for interpreting the isotopic anomalies measured in meteorites. However, the richness of the isotopic measurements forces astrophysicists to continually refine their models.

Key words: **absorption spectra • carbonaceous chondrite • cosmic ray • galaxy • hibonite • interstellar medium • ion probe • ionisation • ionised molecule • isotope • isotopic** (abundance, analysis, anomaly, ratio) **• mass fractionation • mass spectrometer • matrix • molecular cloud • molecular vibration • presolar grain • red giant • refractory inclusion • silicate • solar nebula • spectroscope • star • stellar nucleosynthesis • stellar wind • supernova**

The supernova SN87a, on
February 23, 1987, with the star
Sanduleak before and after the
explosion. A presolar grain of
graphite containing an opaque
grain of titanium carbide.

past two centuries

A few kilograms of meteorite that fell at L'Aigle on April 26th 1803 sparked the scientific study of meteorites. We then had to wait 150 years until the beginning of a rational classification was established.

In 1953, the ages of the Earth, of the solar system and the Universe are still uncertain. The age of the Earth measured by geologists (2.3 billion years or Ga) is even older than the age of the Universe (2 Ga) of the astronomers... Everything changes when the U-Pb chronology of meteorites (1955) fixes the age of the solar system at 4.55 Ga. After that mineralogists, chemists and physicists take meteorites, study them, classify them, extract rare gasses from them, discover the complexity of their components (though some of them, the presolar ones, remain unappreciated another 30 years), detect extinct radioactivities which were alive and well when the solar system formed. Starting in 1969, with the Apollo missions, the emphasis switches for three years to lunar samples. Now the conquest of the Moon did not lead to more fundamental results than those extracted from meteorites since 1973, for a price a hundred times less. We even learn in 1980 that the rise of man could be the eventual result of the impact that seems to have wiped out the dinosaurs!...

If light alone (of every wavelength) is at the root of extraordinary progress in astronomy at the turn of the century, a new laboratory instrument, the ion microscope (or microprobe), allows us to go deeper than telescope observations: it obtains precise chemical and isotopic compositions of remarkable crystals (1/1000 mm) derived from stars born and dead before the Sun was formed. These presolar grains, which have sampled the outer and the deep layers of diverse classes of stars, are realities preserved intact in primitive meteorites.

Autopsied today in the laboratory, their nuclear signatures and intercorrelations often oblige us to modify to modify the relevant astrophysical models. While telescope observations yield average values (stars, stellar atmospheres, molecular clouds, supernova remnants), microscope measurements, grain by grain, extend the spectrum of our observations considerably. Tomorrow, grains from specific comets will be available in the laboratory. Will the two infinities approached by telescopes and microscopes be here and now united?

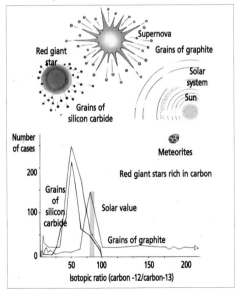

Carbon isotopic ratios in presolar graphite grains from the Murchison meteorite are very diverse, suggesting various origins from stellar nuclear processes and isotopic fractionations. However the same ratios measured for silicon carbide are tightly correlated with the values obtained for red giant stars rich in carbon. We therefore assign its origin to these stars.

Indexed Glossary

Numbers below refer to chapters (Roman numerals, see key in Table of Contents) in which the ideas are used in a significant manner, or to pages (Arabic numbers) which contain a definition of the idea or a detailed explanation of the idea.

Words preceded by an asterisk are defined in other entries in the glossary. Some of the words are defined for this specific context and may be used in a slightly different sense in other fields (e.g. refractory).

ablation: loss of material due to melting and vaporisation during atmospheric passage. 8.

absorption spectrum: spectrum obtained by splitting into its wavelengths the light from a star or an illuminated body, in which light intensity is low at the some wavelengths (absorbtion bonds).

absorption spectrometry: measuring an *absorbtion spectrum to determine a composition.

accretion: process of growth by the accumulation of material from outside. The planets are thus formed by the accumulation of smaller bodies. VII, IX, X.

achondrite: differentiated meteorite made up of silicates. Comes from the crust or mantle of its parent asteroid. V.

aggregate: poorly consolidated accumulation.

albedo: number between 0 and 1, indicating the fraction of luminous energy reflected by an illuminated body. VI.

amino acid: molecular unit incorporated into proteins, containing at least one nitrogen-bearing radical. There are about twenty of them. VIII.

amorphous: solid, without crystalline structure. VII.

angular momentum: a measure of the tendency of a rotating body to continue rotating. Equal to mass times velocity times radius.

asteroid: small planet, mainly rocky and with irregular shape, of very varied size (about 930 km for Ceres, the biggest), orbiting mainly between Mars and Jupiter in a zone called the asteroid belt. The majority of meteorites come from asteroids. 56, 57, VI.

astronomical unit (AU): average distance from the Earth to the Sun (about 150 million km).

axis (of an ellipse): the longest and shortest distances between two points on an ellipse. semi-major axis: half the longer axis, or the longest distance from the "centre" to a point on the ellipse.

basalt: rock resulting from the crystallisation of a lava, composed mainly of plagioclase feldspar (rich in Ca) and pyroxenes. 59.

Big Bang: gigantic explosion thought to generate the Universe. X.

biodiversity: variability of biological species present in an environment. VIII

biosphere: zone of the Earth where animal and plant species live. VIII.

breccia: rock formed by the accumulation of angular fragments of all sizes from one or several rocks. 61, III.

carbonaceous chondrite: *chondrite containing up to several per cent of carbon as organic compounds. V, VIII.

chemical separation: preparation of a rock by which different chemical elements are isolated for individual analysis. 65.

chondrite: un-*differentiated meteorite, largely formed of *chondrules within a matrix. V, (formation) 101.

chondrule: spherule resulting from crystallisation of silicate melt, the accumulation of which within a matrix forms primitive meteorites called *chondrites. 99, V.

circumstellar halo: region of gas and dust rotating around a young star, which will flatten into a disk in a few hundred thousand years. VII.

comet: solar system body consisting of a small ice-rich nucleus which ejects gas and dust when heated by the Sun, forming double tails stretching away from the Sun. 89, VI, VII, VIII.

contamination: mixture with foreign material so as to change physical and chemical properties.

cosmic: coming from space. cosmic material can designate extraterrestrial material in its entirety. Cosmic dust refers to interplanetary grains derived from asteroids and comets.

condensation: formation of a solid or a liquid from a gas. IX.

convection: movement in a liquid or gas, due to spatial variation in temperature.

core (of the Earth): central part of a planet, made up essentially of nickel-iron (at least in the case of the Earth). 57.

cosmic rays: high energy particles (mainly protons). In the solar system, they come from the Sun or from elsewhere in the *galaxy where they have been accelerated by *supernovae. 14.

crater: circular depression excavated by a body, e.g. a very small asteroid, striking a planetary surface. 26, III.

cratering: number of *craters per unit area, considered in relation to time. We speak of the extent of cratering. 33.

crust: outer shell of a planet, especially the earth, in large part *basaltic in composition. The earth's crust is 10-70 km thick. 57.

crystal: organised form of solid matter, in which ions occupy specific sites in a building block (the unit cell), which is repeated indefinitely. V, X.

crystal lattice: arrangement of ions in space in a *crystal with a specific symmetry for each mineral phase.

crystallisation: formation of a *crystal, usually from a liquid. We speak of recrystallisation when there is diffusion of ions and formation of a second generation of crystals in the solid state. V, XI.

dating: measurement of age using a *radioactive chronometer. 105.

differentiation: separation of an initially homogeneous material into several physically and chemically distinct units. 56, 57.

diogenite: *differentiated meteorite containing mainly magnesian pyroxene. Probably a cumulate, i.e. a rock formed by the accumulation of crystals formed in a liquid. 67.

disintegration: transformation of an atomic nucleus into another one, by *radioactive decay. X.

DNA: deoxyribonucleic acid. Complex molecule found in nuclei of cells which carries heredity. VIII.

dust: a small grain or a mass of small grains.

ejecta: material ejected from a crater during an impact. III.

electron microprobe: an instrument using the same principle as a scanning *electron microscope. Characteristic x-rays emitted from micron-sized regions of a sample bombarded by electrons permit chemical analysis. 62.

electron microscope (scanning SEM, transmission TEM): electron accelerator used for very high magnification observations. With an SEM, we observe the diverse types of radiation re-emitted by the surface of the sample to obtain topographic and composition information. 62. With a TEM, diffraction of an electron beam passing through an extremely thin sample gives crystallographic information.

enzymes: proteins able to serve as catalysts, i.e. to accelerate reactions which take place in living cells. VIII.

era: one of the major divisions in geological time. IV.

eucrite: *differentiated meteorite of *basaltic nature. 59, 67, VI.

exposure age: length of a meteorite's interplanetary journey. 14.

extinction: disappearance of all the individuals of a species.

fall: indicates both the motion of a meteorite and, by extension, a meteorite that was observed to fall. Such meteorites are more valuable, because they escape terrestrial alteration and contamination. 1.

fall ellipse: see strewn field.

find: a meteorite found without its fall being observed.

fireball: a very brilliant *meteor. I, II.

focus (of an ellipse): two points inside an ellipse such that the sum of their distances to every point on the ellipse remains constant..

fusion crust: thin glassy skin, usually black, on the surface of a meteorite. 8, V.

galaxy: vast assembly of stars and interstellar matter forming a well defined unit in the Universe, and held together by gravity. By "the Galaxy" we mean the one containing our star (the Milky Way), which contains about 600 billion stars.

glass: an *amorphous solid with an atomic structure like that of a liquid , as opposed to a crystalline solid. Glass is found, with crystalline impact melt rocks, at some craters, and *tektites are glasses caused by impact. Libyan desert glass (99% SiO_2), which has long intrigued scientists, especially Théodore Monod, is now considered by many as having an impact origin.

gravitational anomaly: weight anomaly detected at the Earth's surface due to an inhomogeneous distribution of masses underneath. IV.

half-life (or period): time after which half the *radioactive "parent" *isotope initially present in a reservoir has decayed to the "daughter" isotope. There is a characteristic half-life for each parent-daughter pair. 103, 105, X.

hibonite: refractory mineral, oxide of aluminium and calcium $(CaAl_{12}O_{19})$. XI.

hydrocarbon: organic compound made up of carbon and hydrogen. VIII.

hydrosphere: zone of the Earth consisting of water.

hydrothermal alteration: transformation of minerals in a rock by reactions with a fluid. Refers to water circulating in the parent body, rather than weathering on Earth, after the fall. I, V.

ice: solid (crystalline) form of a normally volatile compound (water ice, methane ice, etc.) 84, 97.

impact: collision between two solid bodies. III.

impact basin: vast but not deep circular structure, resulting from impact of a larger body than one which would make a crater. III.

interstellar: literally "between the stars". Interstellar medium: all the very tenuous material (gas and dust) between the stars in a galaxy. VII, VIII. Interstellar cloud: a relatively dense concentration of matter within the interstellar medium. VII, VIII. Interstellar grain: a grain derived from the interstellar medium. IX.

ion: atom (or molecule) which has lost (or gained) one or several electrons. Because of their positive or negative charge, ions can be accelerated by an electric field and deviated by a magnetic field.

ion probe (or microprobe): an instrument using the same principle as the *mass spectrometer, used to measure *isotopic compositions of samples down to sub-micron size. *Ionisation of the sample is achieved by bombarding with primary *ions so that analysis can be done in-situ (and grains chosen within the textural context of their host rock). 65, 119.

ionisation: loss or gain of electrons from an atom or molecule (which then becomes an *ion).

ionised: ionised matter: whose atoms or molecules are *ions.

iridium: metallic element in the platinum group. IV.

iron: term used for a meteorite made of iron-nickel.

irradiation: bombardment by light photons, electrons, protons, neutrons, etc.

isotope: atomic *nucleus considered as a function of its number of protons and neutrons. 64. If the nucleus is *radioactive, we speak of radio-isotope or parent isotope in relation to the nucleus resulting from the transformation, the daughter isotope. Parent and daughter are isotopes of different chemical elements. V, VII, X, XI.

isotopic: relating to *isotopes. Isotopic abundance: the abundance of a given isotope. Isotopic analysis. 64, 65, 119. Isotopic anomaly: variation in the relative abundance of different isotopes not related to *mass fractionation..

Isotopic composition: relative abundance of different isotopes. Isotopic ratio: to compare isotopic compositions from one sample to another, we express them in the form of a ratio to a reference isotope. 65. Isotopic signature: characteristic isotopic composition.

Kepler's laws: three laws describing the movement of the planets around the sun, stated between 1609 and 1619 by the German astronomer Kepler. VI.

Kirkwood Gaps: discontinuities in the distribution of *asteroid orbits, linked to resonances with Jupiter. 71.

K-T boundary: transition between two geological eras, the Mesozoic, the last period of which was the Cretaceous, and the Tertiary. This boundary is at 65.0 million years ago. IV.

lamellar structure: consisting of lamellae, or plates. In *shocked quartz, there are lamellae of amorphous silica with several different orientations. 43.

light: ordinary, cross-polarised, reflected): these terms describe different conditions for observing rock thin sections in the microscope. With transmitted light (light source behind the thin section) the "ordinary" light is in fact polarised (forced to vibrate in a single pane by passage through a polariser); and (cross-) polarised light is seen after passing through a second polariser at right angles to the first. In reflected light, the illuminated surface serves as a mirror, and the different capabilities of the minerals for reflecting are observed. V.

light year: distance travelled by light in a vacuum in a year, about 9,461 billion km.

magma: Liquid *silicate, resulting from melting of rock.

magnetic minerals: attracting a magnet.

magnetite: iron oxide (Fe_3O_4) containing ions of divalent and trivalent iron.

mantle (of the Earth): intermediate shell of a planet, especially the Earth, made up of rocks rich in *olivine. 57.

mass fractionation: certain physical processes tend to separate the different *isotopes of the same element as a function of their mass. For example, there can be a concentration of "heavy" isotopes in an evaporation residue. The size of the effect for different isotopes depends on their difference in mass from the isotope chosen as a reference. 64, 65, IX.

mass spectrometer: instrument used to measure *isotopic compositions the principle of which is to separate ions of different mass by accelerating them in an electric field and deflecting them with a magnetic field. 64, 65.

matrix: fine-grained material between fragments; in a chondrite between chondrules and refractory inclusions.. 101, V.

metamorphism: transformation of a rock in the solid state (see re*crystallisation), due to increase in temperature and sometimes pressure, e.g. related to burial in parent body. 60.

meteor: luminous phenomenon observed when a body from space passes through the atmosphere. I, II.

meteor stream: group of interplanetary dust particles, in very similar orbits around the Sun, often associated with a periodic *comet. Encounter with Earth usually generates a shower of shooting stars (e.g. the Perseids, which reach a peak on August 12th).

meteorite: a natural object of extraterrestrial origin, which survived passage through the Earth's atmosphere.

micrometeorite: small extraterrestrial particle which survived atmospheric entry, with a size usually much less than a few mm.

molecular cloud: cloud of *interstellar material with gas mainly in the form of molecules (H2). VII, VIII.

molecular vibration: oscillatory movements of the atoms of a molecule with respect to their equilibrium positions.

NEA: Near Earth Asteroids. Asteroids whose orbit brings them close to the Earth's orbit VI.

nebula: cloud of gas and dust having varied origins. "Diffuse" nebulae, such as the Orion nebula, are mainly composed of hydrogen and are star formation regions. VII, VIII.

Neumann bands: network of lines (*twins) in an iron-nickel alloy, seen after light etching, due to mild shock. 61.

Ni-bearing magnetite: *magnetite in which part of the iron ions are replaced by Ni. 44.

nitrogen compound: organic molecule containing one or more atoms of nitrogen, e.g. amino acid. VIII.

nodule: a compact lump, possibly irregular.

nuclear astrophysics: discipline concerned with *nuclear reactions taking place within stars.

nuclear reaction: reaction happening at the level of atomic nuclei (fusion, fission, etc.). 63, XI.

nucleosynthesis: all the processes leading to the appearance of the chemical elements making up matter in the Universe. To a large extent, this implies *nuclear reactions occurring within stars. XI.

nucleus: the extremely tiny heart of an atom. Made up of protons and neutrons, with a positive charge equal to the number of protons.

olivine: iron-magnesium silicate [$(Mg,Fe)_2SiO_4$]. V, IX.

Oort cloud: spherical zone (more than 20,000 AU out) in which *comets reside. 89, VII.

orbit: the trajectory desrcibed by any body around another, e.g. a planet around the Sun, etc. VI.

organic (compound, molecule, substance): containing carbon, hydrogen, oxygen and nitrogen. VIII.

oxidised (element): positively ionised element (usually in association with oxygen). 55

pallasite: *differentiated meteorite consisting of numerous grains of olivine enclosed in iron-nickel metal. 58, 67.

parent body: the body a meteorite comes from. Usually an asteroid, but occasionally the Moon or Mars. VI.

parsec: unit of distance used in astronomy (3.26 light years, or 30,857 billion kilometres).

photosphere: visible region of sun or star like the sun. Most *spectroscopic abundances for the sun were measured in this region. 54, 59.

planet: a non-luminous body orbiting a star, especially the Sun. In the solar system, the term can include bodies of all sizes, including *asteroids. By inner planets we mean the rocky ones closest to the Sun, which are small and dense, with a solid *crust (Mercury, Venus, Earth and Mars, to which the Moon can be added).

planetesimal: small rocky or icy body formed in the solar nebula, in many cases inccorporated into planets. VII, IX, X.

plasma tail: a long straight bluish *comet tail, pointing away from the Sun, made of ions formed in the coma. 92.

plate tectonics: relative movements of the lithospheric plates (*crust and upper rigid part of the *mantle) making up the outer parts of a *differentiated planet. VII.

polymer: organic macromolecule made by repetition of a basic structural unit. VIII.

prebiotic soup: mixture of water and *organic compounds (methane, ammonia...) within which life might have been able to develop. VIII.

presolar grain: a grain formed around another star before the formation of the solar system. XI.

protein: biological macromolecule, made of a chain of amino acids. VIII.

protoplanetary disk: synonym for *solar nebula. IX.

pyrite(s) FeS_2

pyroxene: single-chain silicate containing magnesium, iron and calcium, plus minor elements, in variable quantities. [$(Mg,Fe,Ca)_2Si_2O_6$].

radioactive: see radioactivity. Radioactive chronometer, clock: parent-daughter isotope pair used to determine an age. Radioactive dating: estimating an age using the rate of *radioactive decay of an unstable *isotope. Radioactive element, isotope or nucleus: an unstable atomic nucleus, which will break down due to *radioactivity. VIII, X.

radioactivity: property of certain atomic nuclei of breaking down by emitting particles (especially electrons, neutrinos, gamma rays...) VIII, X.

radiogenic: generated by *radioactive decay of another nucleus. X.

radiometric age: an age determined with a *radioactive chronometer. X.

red giant: large and very luminous star, whose nuclear fuel is almost used up and which will fade in a short time.

reduced: in general, means that an atom is in metallic or neutral form (and not positively charged in association with oxygen). 55.

refractory (element, inclusion): stable at high temperatures under nebular conditions of low pressure. The refractory elements (Al, Ca, Ti...) are the first to condense from a cooling gas (or the last to vaporise from a heated solid). Refractory inclusions are made of minerals rich in refractory elements. V, IX, XI.

regmaglypt: depressions, like fingerprints, on the surface of meteorites produced during atmospheric passage. 10.

regolith: layer of rocky debris produced by impacts on the surface of a planetary body. 36.

resonance: an enhanced response of a system due to an external stimulus with the same (or a simple fraction or multiple of) frequency or period; especially for asteroids with an orbital period which is a simple fraction of the orbital period of Jupiter, leading to a systematically repeated gravitational disturbance. VI.

satellite: a body orbiting a planet (or asteroid).

shatter cone: a conical fracture surface in a rock, formed by the passage of a shock wave during impact of an asteroid. 33.

shock vein: melted material filling a fracture in a rock as a result of *impact. 61.

shock wave: an abrupt and large increase in pressure propagating through a medium.

shocked: having responded to the passage of a shock wave. III, IV.

shocked quartz: quartz (a crystalline form of silica, SiO_2) which experienced *shock. Shocked quartz contains fine lamellae of *amorphous silica. 43.

shooting star: light phenomenon caused by *cosmic dust particle passing through the atmosphere. I.

silica: silicon dioxide.

silicate: mineral containing silicon and oxygen.

solar nebula: disk of gas and dust surrounding the newly formed Sun, in which the planets will develop. IX.

solar prominence: zone of gas blasting off the Sun. 78.

space probe: unmanned vehicle used for study of solar system bodies and interplanetary medium. 80, 90.

spectrograph: instrument for recording light *spectra

using a photographic plate or photoelectric detectors.

spectroscope: optical instrument allowing the observation of a light *spectrum with the eye.

spectrum: the set of monochromatic components resulting when white light or other electromagnetic radiation is split.

star: sphere of gas at very high temperature, within which nuclear fusion reactions occur, making it a source of light and heat. VII, IX, XI. Protostar: star in the process of forming and within which *nuclear reactions are beginning.

stellar wind: continuous flux of charged particles escaping from the atmosphere of certain stars, including the Sun. IX, XI.

strewn field: zone in which fragments of the same meteorite fall are found, usually having the form of an ellipse. The smaller fragments are more easily slowed and travel less far than the big ones: the distribution of fragments as a function of mass yields the direction of the body. 9, 25.

supernova: massive star which reaches an advanced state of evolution and explodes, becoming temporarily very much brighter. 116, 117, VII, XI.

tektite: impact glass formed from surficial material and ejected great distances during the formation of a crater bigger than 10 km in diameter on Earth. 37.

thin section: slice of rock 0.03 mm thick observed in the microscope with transmitted polarised *light.

twin: crystal having domains with different (fixed) crystallographic orientations.

vaporisation: transformation of solid or liquid to the gaseous state. IX.

volatile (element): element occurring as a gas at modest temperatures (e.g. mercury, zinc...). Opposite to *refractory, in that it condenses late from a cooling gas and evaporates early when a liquid or solid is heated. IX.

white inclusion: synonym for *refractory inclusion. V, IX, XI.

Widmanstätten pattern: intergrowth produced by the exsolution of Ni-poor iron alloy from large Ni-rich iron crystals, especially in iron meteorites having more than 6% bulk nickel. 58.

zircon: zirconium silicate ($ZrSiO_4$). Because of their high uranium contents, zircon crystals are often used for dating. 46.

Acknowledgements and Illustration credits

We wish to thank the following, especially for the loan of photos and thin sections:
Altoona Mirror (Pennsylvania); Sachiko Amari (Washington University, St-Louis); American Museum of Naturel History (New York); Margaret E. Bastow (Sky Publishing Corp.); Bibliotheca Apostolica Vaticana; Bibliothèque Centrale of MNHN; Tom Bernatowicz (Washington University, St-Louis); Janet Borg (IAS, Orsay); Alan Boss (DTM, Washington); Martin Beech (Univ. Western Ontario); Cameca; Alain Carion; Ciel et Espace; Harold Connolly (California Inst. of Technology); Ron Cowen (Science News); Jerry Delaney (Rutgers Univ. NJ); Georg Delisle (Geowissenschaften and Rohstoffe); ESA; Marianne Ghélis (MNHN); Rose and Pierre Hewins; Alan Hildebrand (Geol. Survey of Canada); Robert Hutchison (British Mus. Natural History); Institut d'Astrophysique Spatiale; Trevor Ireland (RSES, Australian National Univ.); Jet Propulsion Lab; George McGhee (Rutgers Univ. NJ); Thomas Marvin; Jean-Guy Michard (MNHN); NASA; Florence Raulin-Cerceau (MNHN); Alan Rubin (Univ. California Los Angeles); Smithsonian Astrophysical Observatory, Cambridge, Mass.; Smithsonian Institution (Washington); Staatliche Museen, Berlin; Sophie Tchang (Fondation Hergé); Universitätsbibliothek, Basel; John Wood (Harvard-Smithsonian Center for Astrophysics, Cambridge Mass.).

Photograph credits

Cover: Impact © W. Hartmann/Ciel et Espace; Ida and Dactyle, © NASA; Staunton, © MNHN/Minéralogie/P. Lafaite, thin section, © MNHN/Minéralogie/M. Denise.
p. 6: ph © S. Numazawa/APB/Ciel et Espace
p. 7: (h) ph © A. Carion; (b) ph © S. Eichmiller/Altoona Mirror
p. 10: (h) ph © MNHN/Minéralogie/P. Lafaite; (b, g) ph © MNHN/Minéralogie/G. Segerer; (b, d) ph © MNHN/Minéralogie/L. Bessol/P. Lafaite
p. 12: ph © Mr and Mrs James M. Baker, Lillian, Alabama
p. 13: (b) ph © Smithsonian Astrophysical Observatory
p. 15: (h, g) ph © MNHN, Minéralogie/P. Pellas; (h, d) ph © Hiroshi Matsumoto
p. 16: ph © Private Collection, Fitzwilliam Museum, University of Cambridge
p. 17: (h) ph © U. Marvin; (b) ph © Kupferstichkabinett, Staatliche Museen, Berlin
p. 18: (h) ph © Offentliche Bibliothek, Universitätsbibliothek, Basel; (m) ph © Archives Musée de Minéralogie, Strasbourg; (b) ph © T.C. Marvin
p. 19: (h) ph © Biblioteca Apostolica Vaticana, Rome; (b) ph © MNHN/Bibliothèque centrale
p. 20: 4 ph © MNHN/Bibliothèque centrale
p. 21: ph © MNHN/Minéralogie/P. Lafaite
p. 22: (h) ph © U. Marvin; (b) ph © Science News
p. 23: ph © MNHN/Bibliothèque centrale
p. 24: (h) ph © Margaret E. Bastow, 1995; (b) ph © MNHN/Minéralogie/G. Segerer
p. 25: ph © MNHN/Bibliothèque centrale
p. 26: ph © A. Carion
p. 27: (h) ph © A. Cirou/Ciel et Espace; (b) ph © B. Zanda
p. 28: (g) ph MNHN/P. Lafaite © G. Delisle; (d) ph © NASA
p. 30: ph © O. Hodasawa/NASA/Ciel et Espace
p. 31: ph © LPI/Ciel et Espace
p. 32: 4 ph © NASA
p. 35: (h et b, g) 2 ph © NASA
p. 36: (h) 2 ph © NASA; (b) ph © DR
p. 37: (h) 2 ph © NASA; (b, d) ph © MNHN/Minéralogie/L. Bessol
p. 38: (h) ph © NASA; (b) ph © MNHN/Minéralogie/G. Segerer
p. 39: ph © Palais de la découverte, Paris
p. 40: ph © W. Hartmann/Ciel et Espace
p. 41: 2 ph © MNHN/Paléontologie, D. Serrette
p. 42: (h) 2 ph © R. Rocchia/É. Robin/CEA
p. 43: ph © H. Leroux/UST, Lille
p. 44: (b) ph © É. Robin/CEA
p. 45: (h) ph © S. Numazawa/APB/Ciel et Espace
p. 46: (g) ph © Geological Survey of Canada
p. 47 et 48: 2 ph © MNHN/Paléontologie/D. Serrette
p. 49: (h) ph © MNHN/Paléontologie/D. Serrette; (b) ph © D. Heuclin/Bios
p. 50: ph © MNHN/Minéralogie/G. Segerer
p. 51, 52: 4 ph © MNHN/Minéralogie/P. Lafaite

p. 53: 3 ph © MNHN/Minéralogie/M. Denise
p. 54: (b) 2 ph © MNHN/Minéralogie/C. Caillet-Komorowski
p. 55: (b, g) ph © MNHN/Minéralogie/P. Lafaite; (b, d) ph © MNHN/Minéralogie/M. Denise
p. 56: (h) ph © Sky Publishing Corporation; (b) ph © MNHN/Minéralogie/P. Lafaite
p. 57: (h) ph © Sky Publishing Corporation; (b) ph © E. Graeff/Ciel et Espace
p. 58: (h) 2 ph © MNHN/Minéralogie/P. Lafaite; (b) ph © MNHN/Minéralogie/G. Segerer
p. 59: (m, g) ph © MNHN/Minéralogie/M. Denise; (m, d) ph © MNHN/Minéralogie/M. Christophe; (b) ph © MNHN/Minéralogie
p. 60: 3 ph © MNHN/Minéralogie/M. Denise
p. 61: (h) 2 ph © MNHN/Minéralogie/M. Denise; (b, g) ph © MNHN/Minéralogie/M. Christphe; (b, d) ph © MNHN/Minéralogie/G. Segerer
p. 62-63: 3 ph © MNHN/Minéralogie/M. Denise
p. 68: ph © J.M. Joly/Ciel et Espace
p. 69: ph © NASA
p. 70: (h, d) ph © NASA; (b, d) ph © ESA
p. 71: ph © S. Numazawa/APB/Ciel et Espace
p. 72: (h) ph © ESA/NASA
p. 73: (h) ph © Steven Ostro (GPL); (b) ph © SPL/Ciel et Espace
p. 74: (h) ph © NASA; (b) 2 ph © MNHN/Minéralogie/M. Denise
p. 75: (g) 2 ph © MNHN/Minéralogie/P. Lafaite; (d) ph © DR
p. 76: ph © NASA
p. 77: ph © A. Fujii/Ciel et Espace
p. 78: (g) ph © NASA/Ciel et Espace; (d) ph © NASA
p. 79 et 80: ph © S. Numazawa/APB/Ciel et Espace
p. 81 et 82: 5 ph © NASA
p. 83: (g) ph © NASA; (d) ph © NASA/Ciel et Espace
p. 84: ph © MNHN/Minéralogie/M. Denise
p. 85: ph © J.-P. Bibring/J. Borg/IAS
p. 86: ph © A. Fujii/Ciel et Espace
p. 87: ph © MNHN/Minéralogie
p. 89: (h) ph © ESA/Max-Plank-Institut für Aeronomie (Courtesy H. U. Keller)
p. 90: 2 ph © ESA
p. 91: ph © A. Fujii/Ciel et Espace
p. 92: (b, d) ph © J.-P. Bibring/J. Borg/IAS
p. 93: 2 ph © MNHN/Paléontologie/D. Serrette
p. 94: ph © W.K. Hartmann
p. 95: ph © C. Burrows (ST Scl & ESA), the WFPC 2 Instrument Definition Team, and NASA
p. 96: (b) ph © Manchu/Ciel et Espace
p. 97: (h) 2 ph © MNHN/Minéralogie/M. Denise
p. 98: (h) 3 ph © MNHN/Minéralogie/M. Denise; (m) ph © Harold Connolly; (b) ph © HERGÉ/MOULINSART-1996
p. 99, 101 et 102: 4 ph © MNHN/Minéralogie/M. Denise
p. 106: ph © BEICIP after an image from ESA Darmstadt produced by the Lannion CMD
p. 108: (b) ph © MNHN/Minéralogie/M. Denise
p. 109: text box: (g, h, and d) ph © MNHN/Minéralogie/M. Denise; (g, m): ph © MNHN/Minéralogie/P. Lafaite; (g, b): ph © MNHN/Minéralogie; (b) ph © MNHN/Minéralogie/M. Denise
p. 111: ph © NASA/Ciel et Espace
p. 112: ph © Tom Bernatowicz
p. 116: ph © ESO/Ciel et Espace
p. 117: ph © Jon A. Morse (STScl) and NASA
p. 118: 2 ph © Sachiko Amari
p. 119: ph © CAMECA
p. 120: ph © ESA
p. 122: (background) ph © AAO/D. Malin/Ciel et Espace; (box) ph © Tom Bernatowicz

Drawing credits

Sandra Smith: p. 9 (b), 14, 15 (h), 42 (b), 97 (b)
Bruno Congar: p. 29, 34, 35 (b, d), 38 (m), 44 (h), 45 (h), 57 (m), 62 (h), 64 (h et b), 88, 92 (h), 96 (h), 100, 104, 108 (h), 110, 121, 123
Jean-Marc Trimouille: p. 9 (h), 13 (h), 54 (h), 55 (h), 59 (h), 65, 89 (b), 92 (b, g), 107, 114, 115 (g, adapted from I. R. Littl-Marenin, Astrophysical Journal, 1986, and d)
Jean-Marc Trimouille and Véronique Jara-Ron (MNHN): p. 11, 37 (b, g), 46 (d), 70 (g), 72 (b)